Global Supply Chain

Systems Innovation Series

Series Editor: Adedeji B. Badiru, Air Force Institute of Technology (AFIT) – Dayton, Ohio

Systems Innovation refers to all aspects of developing and deploying new technology, methodology, techniques, and best practices in advancing industrial production and economic development. This entails such topics as product design and development, entrepreneurship, global trade, environmental consciousness, operations and logistics, introduction and management of technology, collaborative system design, and product commercialization. Industrial innovation suggests breaking away from the traditional approaches to industrial production. It encourages the marriage of systems science, management principles, and technology implementation. Particular focus will be the impact of modern technology on industrial development and industrialization approaches, particularly for developing economics. The series will also cover how emerging technologies and entrepreneurship are essential for economic development and society advancement.

Design for Profitability
Guidelines to Cost Effectively Manage the Development Process of Complex Products
Salah Ahmed Mohamed Elmoselhy

Data Analytics
Handbook of Formulas and Techniques
Adedeji B. Badiru

Handbook of Industrial and Systems Engineering
Second Edition
Edited by Adedeji B. Badiru

Innovation Fundamentals
Quantitative and Qualitative Techniques
Adedeji B. Badiru and Gary Lamont

Global Supply Chain
Using Systems Engineering Strategies to Respond to Disruptions
Adedeji B. Badiru

For more information about this series, please visit: https://www.routledge.com/ Systems-Innovation-Book-Series/book-series/CRCSYSINNOV?pd=published, forthcoming&pg=1&pp=12&so=pub&view=list

Global Supply Chain

Using Systems Engineering Strategies to Respond to Disruptions

Adedeji B. Badiru

CRC Press
Taylor & Francis Group
Boca Raton London New York

CRC Press is an imprint of the
Taylor & Francis Group, an **informa** business

First edition published 2022
by CRC Press
6000 Broken Sound Parkway NW, Suite 300, Boca Raton, FL 33487-2742

and by CRC Press
4 Park Square, Milton Park, Abingdon, Oxon, OX14 4RN

ISBN: 978-0-367-63037-9 (hbk)
ISBN: 978-0-367-63039-3 (pbk)
ISBN: 978-1-003-11197-9 (ebk)

DOI: 10.1201/9781003111979

Typeset in Times
by SPi Technologies India Pvt Ltd (Straive)

Dedication

Dedicated to Dr. Sidney G. Gilbreath, Professor Emeritus at Tennessee Technological University, the first industrial engineer to cross my path. As the Director of Materials Management, Eastern Operations at Westmoreland Coal Co. (later renamed Westmoreland Mining LLC) and also as General Manager of the Stonega Division (Virginia Operations), where he led a management team in the reorganization of the Division, in the late 1970s, Dr. Gilbreath inspired me with the diverse applications of industrial engineering to diverse industries. That inspiration is still with me today and served as the motivation for the practical orientation of this book.

Contents

Preface

This book uses a systems-based approach of tools and techniques of industrial engineering applied to the global supply chain. The idea for the book started at the height of the COVID-19 pandemic in early 2000 in response to the growing disruption of the food supply chain. In fact, the initial title of the book was *Food Supply Chain: Using Systems Engineering Strategies for Protection against COVID-19 and Other Pandemics*. It soon became obvious that the need to apply industrial engineering to the supply chain goes beyond the food supply chain, particularly in a pandemic-influenced marketplace. The title, subsequently, changed to *Global Supply Chain: Using Systems Engineering Strategies to respond to Disruptions*, which more appropriately conveys the widespread tentacles of any supply chain.

There are pockets of inefficiency and ineffectiveness in any supply chain, due to a lack of systems accountability. Thus, the focus of this book is on how the diversity and versatility of industrial engineering can be applied to impact process improvement onto a supply chain. It is anticipated that this approach will help achieve better resilience and responsiveness in a supply chain to be more robust against disruptions caused by a pandemic or other market catastrophes.

The book illustrates the applicability of industrial engineering in all facets of endeavors, including business, industry, government, the military, and even academia. Where there is a supply-and-demand issue, there should be an industrial engineering methodology. Topics covered include classical tools and techniques of industrial engineering, general elements of a supply chain, quantitative modeling of a supply chain, efficiency, effectiveness, productivity, systems optimization, enmeshing of qualitative and quantitative techniques of management, innovation in the supply chain, learning curve applications to the supply chain, Triple C principles of communication, cooperation, and coordination, and the application of the DEJI systems model®.

Supply chain strategies, models, and techniques have been around and constantly morphing for decades. This book does not purport to present any magical solution. Rather, it focuses on the presentation of a more comprehensive systems-based approach. The more broadly we can think of the supply chain, from a systems perspective, the more we can have a handle on many of the nuances of the supply chain. The project systems logistics management foundation used in the book facilitates a workable mix of qualitative and quantitative tools and techniques. Readers can expand their knowledge and improve their understanding of the intricacies of globally interconnected supply chains.

Acknowledgments

I thank all my industrial engineering colleagues over the past several decades, who have always inspired me technically, educationally, professionally, and intellectually. I draw my writing inspirations from what I see, hear, and observe in how industrial engineering make diverse contributions to global challenges. Kudos to industrial engineers.

Author

Dr. Adedeji B. Badiru is a professor of Systems Engineering at the Air Force Institute of Technology (AFIT). He is a registered professional engineer and a fellow of the Institute of Industrial Engineers as well as a fellow of the Nigerian Academy of Engineering. He has a BS degree in Industrial Engineering, MS in Mathematics, MS in Industrial Engineering from Tennessee University, and PhD in Industrial Engineering from the University of Central Florida. He is the author of several books and technical journal articles and has received several awards and recognitions for his accomplishments.

1 Project Framework for the Global Supply Chain

THE AMORPHOUS GLOBAL SUPPLY CHAIN

Supply chain requires that the right product, be provided at the right time, in the right place, and in the right quantity. This essentially involves the actualization of what, who, when, where, why, and how (w^4h), in which industrial engineers are very versed because of their training in inquisitive inquiries. Of course, the supply chain can never stop. It is through the tools and techniques of industrial engineering that we can always get to the bottom of supply chain problems. Topics such as KPI (key performance index), predictive statistics, quality certification, engineering economic analysis, IoT (Internet of Things), systems optimization, forecasting, learning curve modeling, and other digital landscapes are within the tool boxes of industrial engineers and can be applied directly to supply chain problems.

The emergence of COVID-19 in 2020 revealed how fragile and shifting the global supply chain can become in the presence of a wide-sweeping disruption. Due to the pandemic, the normally smooth and efficient supply chain does not respond as swiftly and effectively as typically expected. The global supply chain is an interconnected piece of supply routes and linkages. This creates a complex project scenario. Supply chain problems are not always due to product shortages, as consumers often think in their simplistic and uninformed ways. Below are some possible contributing factors, attributes, and indicators that may cause supply chain problems:

- Product shortages
- Transportation problems
- Workforce inequities
- Labor shortage
- Sabotage
- Accidents
- Natural disasters
- Political upheaval
- Hoarding
- Excessive demand vis-à-vis supply

All of these are often recognized in their own individual rights to attention. But when they work in tandem to affect the supply chain, they can spell doom for national operations. The fluidity that may pose problems in the global supply

DOI: 10.1201/9781003111979-1

chain can also turn out to present opportunities for alternate approaches that are responsive and adaptive to the current scenarios, operational needs, and market requirements. That is one justification for calling upon the tools and techniques of industrial engineering. Factory shutdowns in international sources, such as China, and shipping delays at domestic points create production inefficiencies that are best resolved through comprehensive industrial engineering techniques.

Because of the amorphous nature and fluidity of the supply chain, this book uses a triple-helix approach, leveraging industrial engineering, project management, and systems engineering. This calls for a mix of quantitative and qualitative techniques. So, the diverse soft and hard techniques favored by different readers are incorporated into the book.

From a comprehensive project management perspective, a supply chain that is designed to be robust in a steady-state scenario will be more likely to be resilient in a scenario of disruption. This is one of the key messages of this book. Using a project management framework and a systems approach, supply chain can be designed to be more robust, resilient, and responsive in both calm and chaotic scenarios.

INDUSTRIAL ENGINEERING EFFICACY IN THE SUPPLY CHAIN

Due to the multifaceted complexity of the global supply chain, the discipline of industrial engineering is well positioned to bring its tools and techniques to bear on the diverse challenges faced in the supply chain. Industrial engineering, a discipline that has flexibility and diversity of applicability, is defined as articulated below by what an industrial engineer does:

> An industrial engineer is one who is concerned with the design, installation, and improvement of integrated systems of people, materials, information, equipment, and energy by drawing upon specialized knowledge and skills in the mathematical, physical, and social sciences, together with the principles and methods of engineering analysis and design to specify, predict, and evaluate the results to be obtained from such systems.

This is exactly what the global supply chain needs in incorporating the multifaceted nuances of the interconnectedness of supply chains around the world.

BEYOND FOOD SUPPLY

Although food supply is often the most easily noticed disruption of the supply chain, there are many other supply chain problems, including the following:

- Auto parts
- Gasoline
- Computer chips
- Technical workforce

In fact, this book initially started with a focus on food supply chain. It was, subsequently, expanded to general supply chains due to the other realities that emerged as a result of the pandemic and other disruptive events. In most cases, the supply chain is characterized by the following elements:

- Prediction (to forecast market needs)
- Production (to meet the forecasted demand)
- Transportation (to get the goods out to the market)
- Distribution (to get the goods to the points of sale)
- Transaction (selling of the goods through wholesale, retail, etc.)
- Consumption (buyer's usage of the goods)
- Repetition (repeat of the entire process to keep the market going)

This cycle pertains not only to physical goods but also to the provision of services, such as in healthcare, travel, security, emergency response, and recreation. In fact, the movement of people through the airport is a form of a supply chain process and the cycle described above is directly applicable. Askin and Sefair (2021) present an analytical modeling technique that fits this supposition perfectly. Their work involves improving the efficiency of airport security screening checkpoints. The first step in matching capacity to demand is to predict demand, that is the forecasting of the number of passengers arriving in each time segment. While this may sound simple, the reality is more complicated due to data availability and inherent randomness. Any disruption in the whole transportation system involving a particular airport could easily upend the passenger flow at the local airport level. For a practical implementation of their approach, Askin and Sefair (2021) developed a quantitative model of the number of passengers arriving at security screening checkpoints for specific time intervals on various days. Other aspects of the methodology involve estimating queue lengths and wait times. This is a stand component of the application of operations research to problem scenarios. The work was done at the research Center for Accelerating Operational Efficiency (CAOE) at Arizona State University.

EXPORT–IMPORT COMPLEXITY OF A SUPPLY CHAIN

One case example of the complicated global supply chain can be seen in the seafood industry. It is reported that 80% of the seafood consumed in the United States of America is imported (Zomorodi and Geiran, 2021; NOAA, 2021). In the same breath, 25% of the seafood caught, raised, or farmed in the United States of America is sent overseas for processing before being imported back into the USA market. This symbiotic but amorphous import–export relationship can create challenges even in steady-state times, not to talk of disrupted market times. With the increasing interest in healthier living, consumers are turning more and more to seafood. Thus, the global seafood demand and supply will continue to grow. Trade logistics and agreements may shift, but the basic uptick of supply and demand will continue. Global trade tensions, political uncertainties, pandemic-caused economic downturns, and social unrests may disrupt standard supply chain practices.

However, new technological breakthroughs in aquaculture techniques and biosecurity may offer enhanced responses. A system view of the overall market landscape is essential for being responsive and adaptive to the changing demand and supply profiles.

INPUT-PROCESS-OUTPUT REPRESENTATION

Even for a simple market commodity, the supply chain is a complex chain of input, process, and output as depicted in the Input, Constraints, Outputs, and Mechanisms (ICOM) model shown in Figure 1.1.

INPUT

The supply chain starts with the market specifying the needs, requirements, conditions, and expectations. This is the input phase. Without knowing the nuances of the inputs, the supply chain will start out on the wrong foot. Even little things can matter bigly in a supply chain. This is akin to the age-old proverb by Benjamin Franklin as echoed below:

For the want of a nail the shoe was lost,
For the want of a shoe the horse was lost,
For the want of a horse the rider was lost,
For the want of a rider the battle was lost,
For the want of a battle the kingdom was lost,
And all for the want of a horseshoe-nail.

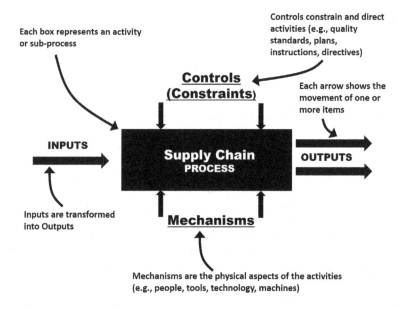

FIGURE 1.1   ICOM Model for the Supply Chain.

From a systems perspective, all inputs of the market are important. It is in the input segment that the supply chain objectives and expected results are specified. Like the scope of work in project management, the scope of the supply chain market also is important to be specified as a part of the inputs.

CONSTRAINTS

There is no flawless or perfect system anywhere. Every supply chain will be subject to constraints, both internal and external. Constraints are the reality of any operation. Problems and impediments crop up and must be addressed promptly before disastrous consequences can develop. Some potential constraints include budget limitation, financial requirements, reporting expectations, schedule impositions, legal guidelines, contractual requirements, trade embargoes, environmental issues, quality standards, and government regulations. Using a systems approach, the linkages between the constraints can be better understood and managed. Supply chain objectives may have to be modified in response to the prevailing constraints. Multi-constrained operation poses a challenge. But a mix of qualitative and quantitative tools can help resolve, overcome, or mitigate the challenges.

MECHANISM

In a supply chain, the mechanism are the people, knowledge base, tools, techniques, models, expertise, talents, technology, and capital. These constitute the "engine" required to move the supply chain forward. The list of mechanisms facilitating the forward movement of the supply chain will often be dynamic. The operators must be responsive and adaptive to adding new mechanisms. For example, the emergence of the COVID-19 pandemic necessitated adding new mechanisms (e.g., COVID testing and vaccination) to several supply chains.

OUTPUT

The output is like the "coin of the realm" of the supply chain enterprise. The output is the valued result of all the inner workings of the supply chain. Outputs can be defined in terms of the generation of a physical product, the provision of a service, or the achievement of a desirable result. Market surveys are important for accurately assessing market expectations and ascertaining the quality of the output. On implementation, the ICOM model provides guidance for specifying Key Result Areas (KRAs) and Key Performance Indicators (KPIs) which can guide the success and resilience of the supply chain.

The input-process-output framework of the ICOM model can be seen and felt at the gasoline pumps in local communities. Even when there is plenty of gasoline in the nation, getting it to the fuel pumps at gas stations is a different story. There are often reports of some stations running out of gasoline for short periods. The problem is often traced to a shortage of tanker truck drivers to transport the fuel from storage terminals to the fuel stations. From the consumer's perspective, the fuel needs to be where it is needed. But if the fuel is plentiful without truck drivers to drive the trucks to move the fuel to its respective points of sale, the fuel will be a "no-show" at the local level. Even shortages of short periods can create annoyance, inconvenience, and predicament that can adversely affect human behavior, leading to events such as a road rage. The whole thing is a complex system with several intervening layers of needs and consideration. This again affirms the needs of a global systems view of the supply chain.

Apart from tanker trucks moving gasoline from point to point, there are other elements in the global supply chain that are not often physically seen at the local level. For example, during the COVID-19 pandemic, reports circulated about merchant ship crews stuck at sea or in ports. If there is no movement of international seafaring ships due to the pandemic, economic issues, trade wars, or other reasons, there will be no supply of goods to points in the global network. It is well known how computer chip shortages affect not only the computer production industry but also the automobile industry and several feeder industries. We can all feel the pains of problems in the global supply chain.

PROJECT SYSTEMS FOR THE SUPPLY CHAIN

It is a systems world in the long run. Things work better with a systems approach. In the final analysis, every endeavor is a project. Applying an integration of the principles of systems management and project management provides a more robust handle of the overall supply chain. Thus, the rest of this chapter discusses various aspects of a project systems framework for the supply chain.

A systems view of a project makes the project execution more agile, efficient, and effective. A system is a collection of interrelated elements working together synergistically to achieve a set of objectives. Any project is a collection of interrelated activities, people, tools, resources, processes, and other assets brought together in the pursuit of a common goal. The goal may be in terms of generating a physical product, providing a service, or achieving a specific result. This makes it possible to view any project as a system that is amenable to all the classical and modern concepts of systems management.

Project management is the foundation of everything we do. Having a knowledge is not enough, we must apply the knowledge strategically and systematically for it to be of any use. The knowledge must be applied to do something in the pursuit of objectives. Project management facilitates the application of knowledge and willingness to actually accomplish tasks. Where there is knowledge, willingness often follows. But it is the project execution that actually gets jobs accomplished. From the very basic tasks to the very complex endeavors, project management must be applied to get things done. It is, thus, essential that project management be a part of the core of every supply chain improvement initiative. The tools and techniques presented here are applicable to any project-oriented pursuit not only a supply chain but also everywhere else. Everything that everyone does can be defined as a project. In this regard, a systems approach is of utmost importance in the global supply chain. Systems application encompasses the following elements:

- Technological systems (e.g., engineering control systems and mechanical systems)
- Organizational systems (e.g., work process design and operating structures)
- Human infrastructure (e.g., workforce development, interpersonal relationships, human-systems integration)

In systems-based project management, it is essential that related techniques be employed in an integrated fashion so as to maximize the total project output. One definition of systems project management offered here is stated as follows:

Systems project management is the process of using systems approach to manage, allocate, and time resources to achieve systems-wide goals in an efficient and expeditious manner.

The definition calls for a systematic integration of technology, human resources, and work process design to achieve goals and objectives. There should be a balance in the synergistic integration of humans and technology. There should not be an overreliance on technology nor should there be an overdependence on human processes. Similarly, there should not be too much emphasis on analytical models to the detriment of common-sense human-based decisions. Particularly in this era of digital operations, many cyber risks exist to upend even the best designed systems. Recent hacking incidents around the world point to the need to exercise all human-based caution to mitigate digital-based risks.

Engineering problem-solving methodology can boost the readiness and resilience of supply chains. In this regard, the techniques of systems engineering have proven effective in addressing multifaceted challenges, even in nonengineering platforms. Systems engineering is growing in appeal as an avenue to achieve organizational goals and improve operational effectiveness and efficiency. Researchers and practitioners in business, industry, and government are collaboratively embracing systems engineering implementations. Several definitions of systems engineering exist. The following is one comprehensive definition:

Systems engineering is the application of engineering to solutions of a multifaceted problem through a systematic collection and integration of parts of the problem with respect to the life cycle of the problem. It is the branch of engineering concerned with the development, implementation, and use of large or complex systems. It focuses on specific goals of a system considering the specifications, prevailing constraints, expected services, possible behaviors, and structure of the system. It also involves a consideration of the activities required to assure that the system's performance matches the stated goals. Systems engineering addresses the integration of tools, people, and processes required to achieve a cost-effective and timely operation of the system.

LOGISTICS IN PROJECT SYSTEMS

Logistics can be defined as the planning and implementation of a complex task, the planning and control of the flow of goods and materials through an organization or manufacturing process, or the planning and organization of the movement of personnel, equipment, and supplies. Complex projects represent a hierarchical system of operations. Thus, we can view a project system as collection of interrelated projects all serving a common end goal. Consequently, the following universal definition is applicable to supply chains:

> Project systems logistics is the planning, implementation, movement, scheduling, and control of people, equipment, goods, materials, and supplies across the interfacing boundaries of several related projects.

Conventional project management must be modified and expanded to address the unique requirements for logistics in project systems.

SUPPLY CHAIN SYSTEMS CONSTRAINTS

Any supply chain is a complex system with multiple constraints. The earlier the prevailing constraints are recognized and attended to the better for a smoother operation later on. Systems management is the pursuit of organizational goals within the constraints of time, cost, and quality expectations. The iron triangle, often referred to as the triple constraints, shows that project accomplishments are constrained by the boundaries of quality, time, and cost. In this case, quality represents the composite collection of project requirements. In a situation where precise optimization is not possible, there will have to be trade-offs between these three factors of success. The concept of iron triangle is that a rigid triangle of constraints encases the project. Everything must be accomplished within the boundaries of time, cost, and quality. If a better quality is expected, a compromise along the axes of time and cost must be executed, thereby altering the shape or profile of the iron triangle.

The trade-off relationships are not linear and must be visualized in a multidimensional context. This is better articulated by a three-dimensional view of the systems constraints using applicable schematics of the supply chain characteristics. Scope requirements determine the project boundary, and trade-offs must be done within that boundary.

SYSTEMS INFLUENCE THEORY

Systems influence philosophy suggests the realization that you control the internal environment while only influencing the external environment. The internal (controllable) environment is represented as a black box in the typical input–process–output relationship. The outside (uncontrollable) environment is bounded by a cloud representation for the unknowns. In the comprehensive systems structure, inputs come from the global environment, are moderated by the immediate outside environment, and are delivered to the inside environment. In an unstructured internal environment, functions occur as blobs. A "blobby" environment is characterized by intractable activities where everyone is busy, but without a cohesive structure of input–output relationships. In such a case, the following disadvantages may be present:

- Lack of traceability
- Lack of process control
- Higher operating cost
- Inefficient personnel interfaces
- Unrealized technology potentials

Organizations often inadvertently fall into the blobs structure because it is simple, low cost, and less time-consuming until a problem develops. A desired alternative is to model the project system using a systems value-stream structure, where each point-to-point interface is identifiable, traceable, and controllable, particularly if the project is implemented in choreographed increments. This suggests using a proactive and problem-preempting approach to execute projects. This alternative has the following advantages:

- Problem diagnosis is easier
- Accountability is higher
- Operating waste is minimized
- Conflict resolution is faster
- Value points are traceable

SUPPLY CHAIN MANAGEMENT BY PROJECT

Project management continues to grow as an effective means of managing functions of any form in any type of organization. Project management should be an enterprise-wide systems-based endeavor. Enterprise-wide project management is the application of project management techniques and practices across the full spectrum of the enterprise. This concept is also referred to as management by project (MBP). MBP is an effective approach that employs project management techniques in various functions within an organization. MBP recommends pursuing endeavors as project-oriented activities. In this respect, every endeavor, large or small, simple or complex, can be modeled as a project and managed rigorously accordingly. Project management is an effective way to conduct any business activity. It represents a disciplined approach that defines any work assignment as a project. Under MBP, every undertaking is viewed as a project that must be managed just like a traditional project.

The traditional definition of a project as "a unique one-of-a-kind endeavor that has a definite beginning and a definite end" is still applicable in the sense that we can cobble together a series of such "definitely-bounded" projects to achieve a composite project that spans multiple time periods and multiple end products.

This was shown by Badiru (1988) in the novel application of project management to manufacturing, which was, hitherto, considered not to be a traditional project. The characteristics required of each project so defined are:

1. An identified scope and a goal
2. A desired completion time
3. Availability of resources
4. A defined performance measure
5. A measurement scale for review of work

An MBP approach to operations helps in identifying unique entities within functional requirements. This identification helps determine where functions overlap and how they are interrelated, thus paving the way for better planning, scheduling, and control. Enterprise-wide project management facilitates a unified view of organizational goals and provides a way for project teams to use information generated by other departments to carry out their functions.

The use of project management continues to grow rapidly. The need to develop effective management tools increases with the increasing complexity of new technologies and processes. The life cycle of a new product to be introduced into a competitive market is a good example of a complex process that must be managed with integrative project management approaches. The product will encounter management functions as it goes from one stage to the next. Project management will be needed throughout the design and production stages of the product. Project management will be needed in developing marketing, transportation, and delivery strategies for the product. When the product finally gets to the customer, project management will be needed to integrate its use with those of other products within the customer's organization.

The need for a project management approach is established by the fact that a project will always tend to increase in size even if its scope is narrowing. The following three literary laws are applicable to any project environment:

Parkinson's law: Work expands to fill the available time or space.
Peter's principle: People rise to the level of their incompetence.
Murphy's law: Whatever can go wrong will.
Badiru's rule: The grass is always greener where you most need it to be dead.

The COVID-19 pandemic confirmed that things can go wrong when we least expect. So, extra precautions are necessary. An integrated systems project management approach can help diminish the adverse impacts of these laws through good project planning, contingency readiness, organizing, scheduling, and control. Project management tools can be classified into three major categories:

1. *Qualitative tools*: There are managerial tools that aid in the interpersonal and organizational processes required for project management.

2. *Quantitative tools*: These are analytical techniques that aid in the computational aspects of project management.
3. *Computer tools*: These are software and hardware tools that simplify the process of planning, organizing, scheduling, and controlling a project. Software tools can help in both qualitative and quantitative analyses needed for project management.

Although individual books dealing with management principles, optimization models, and computer tools are available, there are a few guidelines for the integration of the three areas for project management purposes. In this book, we integrate these three areas for a comprehensive guide to project management. The book introduces the *triad approach* to improve the effectiveness of project management with respect to schedule, cost, and performance constraints within the context of systems modeling. The approach considers not only the management of the project itself but also the management of all the functions that support the project. It is one thing to have a quantitative model, but it is a different thing to be able to apply the model to real-world problems in a practical form. The systems approach presented in this book illustrates how to make the transition from model to practice.

A systems approach helps increase the intersection of the three categories of project management tools and, hence, improve overall management effectiveness. Crisis should not be the instigator for the use of project management techniques. Project management approaches should be used upfront to prevent avoidable problems rather than to fight them when they develop. What is worth doing is worth doing well, right from the beginning.

CRITICAL SUCCESS FACTORS FOR SUPPLY CHAIN PROJECTS

The premise of this section is that the critical factors for systems success revolve around people and the personal commitment and dedication of each person. No matter how good a technology is and no matter how enhanced a process might be, it is ultimately the people involved that determine success. This makes it imperative to take care of people issues first in the overall systems approach to project management. Many organizations recognize this, but only few have been able to actualize the ideals of managing people productively. Execution of operational strategies requires forthrightness, openness, and commitment to get things done. Lip service and arm waving are not sufficient. Tangible programs that cater to the needs of people must be implemented. It is essential to provide incentives, encouragement, and empowerment for people to be self-actuating in determining how best to accomplish their job functions. A summary of critical factors for systems success encompasses the following:

- Total system management (hardware, software, and people)
- Operational effectiveness
- Operational efficiency

- System suitability
- System resilience
- System affordability
- System supportability
- System life cycle cost
- System performance
- System schedule
- System cost

Systems engineering tools, techniques, and processes are essential for project life cycle management to make goals possible within the context of *SMART* principles, which are represented as follows:

1. *Specific*: Pursue specific and explicit outputs.
2. *Measurable*: Design outputs that can be tracked, measured, and assessed.
3. *Achievable*: Pursue outputs that are achievable and aligned with organizational goals.
4. *Realistic*: Pursue only the goals that are realistic and result oriented.
5. *Timed*: Make outputs timed to facilitate accountability.

PMBOK FRAMEWORK FOR SUPPLY CHAIN

The general project management body of knowledge (PMBOK®) was developed by the project management institute (PMI). The body of knowledge comprises specific knowledge areas, which are organized into specific broad areas, all of which are applicable to a supply chain network.

1. Project *integration* management
2. Project *scope* management
3. Project *time* management
4. Project *cost* management
5. Project *quality* management
6. Project *human resource* management
7. Project *communications* management
8. Project *risk* management
9. Project *procurement and subcontract* management

The listed segments of the body of knowledge of project management cover the range of functions associated with any project, particularly complex ones. Multinational projects particularly pose unique challenges pertaining to reliable power supply, efficient communication systems, credible government support, dependable procurement processes, consistent availability of technology, progressive industrial climate, trustworthy risk mitigation infrastructure, regular supply of skilled labor, uniform focus on quality of work, global consciousness, hassle-free bureaucratic processes, coherent safety and security system, steady law and order, unflinching focus on customer satisfaction, and fair labor relations. Assessing and resolving concerns about these issues in a step-by-step fashion will create a foundation of success for a large project. While no system can be perfect and satisfactory in all aspects, a tolerable trade-off on the factors is essential for project success.

The key components of each element of the body of knowledge are summarized as follows:

Integration management
    Integrative project charter
    Project scope statement
    Project management plan
    Project execution management
    Change control
Scope management
    Focused scope statements
    Cost/benefits analysis
    Project constraints
    Work breakdown structure
    Responsibility breakdown structure
    Change control
Time management
    Schedule planning and control
    PERT and Gantt charts
    Critical path method (CPM)
    Network models
    Resource loading
    Reporting
Cost management
    Financial analysis
    Cost estimating
    Forecasting
    Cost control
    Cost reporting
Quality management
    Total quality management
    Quality assurance
    Quality control
    Cost of quality
    Quality conformance
Human resource management
    Leadership skill development
    Team building
    Motivation

Conflict management
Compensation
Organizational structures
Communications management
Communication matrix
Communication vehicles
Listening and presenting skills
Communication barriers and facilitators
Risk management
Risk identification
Risk analysis
Risk mitigation
Contingency planning
Procurement and subcontract management
Material selection
Vendor prequalification
Contract types
Contract risk assessment
Contract negotiation
Contract change orders

It should be noted that project life cycle is distinguished from product life cycle. Project life cycle does not explicitly address operational issues whereas product life cycle is mostly about operational issues starting from the product's delivery to the end of its useful life. Note that for supply chain projects, the shape of the life cycle curve may be expedited due to the rapid developments that often occur in digital operations. For example, for a high technology project, the entire life cycle may be shortened, with a very rapid initial phase, even though the conceptualization stage may be very long. Typical characteristics of project life cycle include the following:

1. Cost and staffing requirements are lowest at the beginning of the project and ramp up during the initial and development stages.
2. The probability of successfully completing the project is lowest at the beginning and highest at the end. This is because many unknowns (risks and uncertainties) exist at the beginning of the project. As the project nears its end, there are fewer opportunities for risks and uncertainties.
3. The risks to the project organization (project owner) are lowest at the beginning and highest at the end. This is because not much investment has gone into the project at the beginning, whereas much has been committed by the end of the project. There is a higher sunk cost manifested at the end of the project.
4. The ability of the stakeholders to influence the final project outcome (cost, quality, and schedule) is highest at the beginning and gets progressively lower toward the end of the project. This is intuitive because influence is best exerted at the beginning of an endeavor.

5. Value of scope changes decreases over time during the project life cycle, while the cost of scope changes increases over time. The suggestion is to decide and finalize scope as early as possible. If there are to be scope changes, do them as early as possible.

SUPPLY CHAIN PROJECT STRUCTURE

PROBLEM IDENTIFICATION

Problem identification is the stage where a need for a proposed project is identified, defined, and justified. A project may be concerned with the development of new products, implementation of new processes, or improvement of existing facilities.

PROJECT DEFINITION

Project definition is the phase at which the purpose of the project is clarified. A *mission statement* is the major output of this stage. For example, a prevailing low level of productivity may indicate a need for a new manufacturing technology. In general, the definition should specify how project management may be used to avoid missed deadlines, poor scheduling, inadequate resource allocation, lack of coordination, poor quality, and conflicting priorities.

PROJECT PLANNING

A plan represents the outline of the series of actions needed to accomplish a goal. Project planning determines how to initiate a project and execute its objectives. It may be a simple statement of a project goal or it may be a detailed account of procedures to be followed during the project. Planning can be summarized as:

Objectives
Project definition
Team organization
Performance criteria (time, cost, and quality)

PROJECT ORGANIZING

Project organization specifies how to integrate the functions of the personnel involved in a project. Organizing is usually done concurrently with project planning. Directing is an important aspect of project organization. Directing involves guiding and supervising the project personnel. It is a crucial aspect of the management function. Directing requires skillful managers who can interact with subordinates effectively through good communication and motivation techniques. A good project manager will facilitate project success by directing his or her staff, through proper task assignments, toward the project goal.

Workers perform better when there are clearly defined expectations. They need to know how their job functions contribute to the overall goals of the project. Workers should be given some flexibility for self-direction in performing their functions. Individual worker needs and limitations should be recognized by the manager when directing project functions. Directing a project requires skills dealing with motivating, supervising, and delegating.

RESOURCE ALLOCATION

Project goals and objectives are accomplished by allocating resources to functional requirements. Resources can consist of money, people, equipment, tools, facilities, information, skills, and so on. These are usually in short supply. The people needed for a particular task may be committed to other ongoing projects. A crucial piece of equipment may be under the control of another team.

PROJECT SCHEDULING

Timeliness is the essence of project management. Scheduling is often the major focus in project management. The main purpose of scheduling is to allocate resources so that the overall project objectives are achieved within a reasonable time span. Project objectives are generally conflicting in nature. For example, minimization of the project completion time and minimization of the project cost are conflicting objectives. That is, one objective is improved at the expense of worsening the other objective. Therefore, project scheduling is a multiple-objective decision-making problem.

In general, scheduling involves the assignment of time periods to specific tasks within the work schedule. Resource availability, time limitations, urgency level, required performance level, precedence requirements, work priorities, technical constraints, and other factors complicate the scheduling process. Thus, the assignment of a time slot to a task does not necessarily ensure that the task will be performed satisfactorily in accordance with the schedule. Consequently, careful control must be developed and maintained throughout the project scheduling process.

PROJECT TRACKING AND REPORTING

This phase involves checking whether or not project results conform to project plans and performance specifications. Tracking and reporting are prerequisites for project control. A properly organized report of the project status will help identify any deficiencies in the progress of the project and help pinpoint corrective actions.

PROJECT CONTROL

Project control requires that appropriate actions be taken to correct unacceptable deviations from expected performance. Control is actuated through measurement, evaluation, and corrective action. Measurement is the process of measuring the relationship between planned performance and actual performance with respect to project objectives. The variables to be measured, the measurement scales, and the measuring approaches should be clearly specified during the planning stage. Corrective actions may involve rescheduling, reallocation of resources, or expedition of task performance.

Tracking and reporting
Measurement and evaluation
Corrective action (plan revision, rescheduling, updating)

PROJECT TERMINATION

Termination is the last stage of a project. The phaseout of a project is as important as its initiation. The termination of a project should be implemented expeditiously. A project should not be allowed to drag on after the expected completion time. A terminal activity should be defined for a project during the planning phase. An example of a terminal activity may be the submission of a final report, the power on of new equipment, or the signing of a release order. The conclusion of such an activity should be viewed as the completion of the project. Arrangements may be made for follow-up activities that may improve or extend the outcome of the project. These follow-up or spin-off projects should be managed as new projects but with proper input–output relationships within the sequence of projects.

TRADITIONAL PROJECT IMPLEMENTATION TEMPLATE

While this chapter advocates the main tenets of the PMI's PMBOK, it also recommends the traditional project management framework encompassing the broad sequence summarized below:

Planning → Organizing → Scheduling → Control → Termination

An outline of the functions to be carried out during a project should be made during the planning stage of the project. A model for such an outline is presented hereafter. It may be necessary to rearrange the contents of the outline to fit the specific needs of a project.

PLANNING

1. Specify project background
   a. Define current situation and process
      i. Understand the process
      ii. Identify important variables
      iii. Quantify variables

   b. Identify areas for improvement
       i. List and discuss the areas
       ii. Study potential strategy for solution
2. Define unique terminologies relevant to the project
   a. Industry-specific terminologies
   b. Company-specific terminologies
   c. Project-specific terminologies
3. Define project goal and objectives
   a. Write mission statement
   b. Solicit inputs and ideas from personnel
4. Establish performance standards
   a. Schedule
   b. Performance
   c. Cost
5. Conduct formal project feasibility study
   a. Determine impact on cost
   b. Determine impact on organization
   c. Determine project deliverables
6. Secure management support

ORGANIZING

1. Identify project management team
   a. Specify project organization structure
       i. Matrix structure
       ii. Formal and informal structures
       iii. Justify structure
   b. Specify departments involved and key personnel
       i. Purchasing
       ii. Materials management
       iii. Engineering, design, manufacturing, and so on
   c. Define project management responsibilities
       i. Select project manager
       ii. Write project charter
       iii. Establish project policies and procedures
2. Implement triple C model
   a. Communication
       i. Determine communication interfaces
       ii. Develop communication matrix
   b. Cooperation
       i. Outline cooperation requirements, policies, and procedures
   c. Coordination
       i. Develop work breakdown structure
       ii. Assign task responsibilities
       iii. Develop responsibility chart

SCHEDULING (RESOURCE ALLOCATION)

1. Develop master schedule
   a. Estimate task duration
   b. Identify task precedence requirements
      i. Technical precedence
      ii. Resource-imposed precedence
      iii. Procedural precedence
   c. Use analytical models
      i. CPM
      ii. PERT
      iii. Gantt chart
      iv. Optimization models

CONTROL (TRACKING, REPORTING, AND CORRECTION)

1. Establish guidelines for tracking, reporting, and control
   a. Define data requirements
      i. Data categories
      ii. Data characterization
      iii. Measurement scales
   b. Develop data documentation
      i. Data update requirements
      ii. Data quality control
      iii. Establish data security measures
2. Categorize control points
   a. Schedule audit
      i. Activity network and Gantt charts
      ii. Milestones
      iii. Delivery schedule
   b. Performance audit
      i. Employee performance
      ii. Product quality
   c. Cost audit
      i. Cost containment measures
      ii. Percent completion versus budget depletion
3. Identify implementation process
   a. Comparison with targeted schedules
   b. Corrective course of action
      i. Rescheduling
      ii. Reallocation of resources

TERMINATION (CLOSE, PHASEOUT)

1. Conduct performance review
2. Develop strategy for follow-up projects
3. Arrange for personnel retention, release, and reassignment

1. Document project outcome
2. Submit final report
3. Archive report for future reference

VALUE OF LEAN OPERATIONS

Facing a lean period in supply chain project management creates value in terms of figuring out how to eliminate or reduce operational waste that is inherent in many human-governed processes. It is a natural fact that having to make do with limited resources creates opportunities for resourcefulness and innovation, which requires an integrated-systems view of what is available and what can be leveraged. The lean principles that are embraced by business, industry, and government have been around for a long time. It is just that we are now being forced to implement lean practices in diverse operational challenges due to the escalating shortage of resources. It is unrealistic to expect that problems that have enrooted themselves in different parts of an organization can be solved by a single-point attack. Rather, a systematic probing of all the nooks and corners of the problem must be assessed and tackled in an integrated manner.

Contrary to the contention in some technocratic circles that budget cuts will stifle innovation, it is a fact that a reduction of resources often forces more creativity in identifying wastes and leveraging opportunities that lie fallow in nooks and crannies of an organization. This is not an issue of wanting more for less. Rather, it is an issue of doing more with less. It is through a systems viewpoint that new opportunities to innovate can be spotted. Necessity and adversity can, indeed, spark invention.

SUPPLY CHAIN DECISION ANALYSIS

Systems decision analysis facilitates a proper consideration of the essential elements of decisions in a project systems environment. These essential elements include the problem statement, information, performance measure, decision model, and an implementation of the decision. The recommended steps are enumerated as follows:

STEP 1. PROBLEM STATEMENT

A problem involves choosing between competing, and probably conflicting, alternatives. The components of problem solving in project management include:

Describing the problem (goals, performance measures)
Defining a model to represent the problem
Solving the model
Testing the solution
Implementing and maintaining the solution

Problem definition is very crucial. In many cases, *symptoms* of a problem are more readily recognized than its *cause* and *location*. Even after the problem is accurately identified and defined, a benefit/cost analysis may be needed to determine if the cost of solving the problem is justified.

Step 2. Data and Information Requirements

Information is the driving force for the project decision process. Information clarifies the relative states of past, present, and future events. The collection, storage, retrieval, organization, and processing of raw date are important components for generating information. Without data, there can be no information. Without good information, there cannot be a valid decision. The essential requirements for generating information are:

Ensuring that an effective data collection procedure is followed
Determining the type and the appropriate amount of data to collect
Evaluating the data collected with respect to information potential
Evaluating the cost of collecting the required data

For example, suppose a manager is presented with a recorded fact that says, "Sales for the last quarter are 10,000 units." This constitutes ordinary data. There are many ways of using the aforementioned data to make a decision, depending on the manager's value system. An analyst, however, can ensure the proper use of the data by transforming it into information, such as "Sales of 10,000 units for the last quarter are within x percent of the targeted value." This type of information is more useful to the manager for decision-making.

Step 3. Performance Measure

A performance measure for the competing alternatives should be specified. The decision-maker assigns a perceived worth or value to the available alternatives. Setting measures of performance is crucial to the process of defining and selecting alternatives. Some performance measures, commonly used in project management, are project cost, completion time, resource usage, and stability in the workforce.

Step 4. Decision Model

A decision model provides the basis for the analysis and synthesis of information and is the mechanism by which competing alternatives are compared. To be effective, a decision model must be based on a systematic and logical framework for guiding project decisions. A decision model can be a verbal, graphical, or mathematical representation of the ideas in the decision-making process. A project decision model should have the following characteristics:

Simplified representation of the actual situation
Explanation and prediction of the actual situation

Validity and appropriateness
Applicability to similar problems

The formulation of a decision model involves three essential components:

Abstraction: Determining the relevant factors
Construction: Combining the factors into a logical model
Validation: Assuring that the model adequately represents the problem

The basic types of decision models for project management are described next:

Descriptive models: These models are directed at describing a decision scenario and identifying the associated problem. For example, a project analyst might use a CPM network model to identify bottleneck tasks in a project.

Prescriptive models: These models furnish procedural guidelines for implementing actions. The Triple C approach (Badiru, 2019a), for example, is a model that prescribes the procedures for achieving communication, cooperation, and coordination in a project environment.

Predictive models: These models are used to predict future events in a problem environment. They are typically based on historical data about the problem situation. For example, a regression model based on past data may be used to predict future productivity gains associated with expected levels of resource allocation. Simulation models can be used when uncertainties exist in the task durations or resource requirements.

Satisficing models: These are models that provide trade-off strategies for achieving a satisfactory solution to a problem, within given constraints. Goal programming and other multi-criteria techniques provide good satisficing solutions. For example, these models are helpful in cases where time limitations, resource shortages, and performance requirements constrain the implementation of a project.

Optimization models: These models are designed to find the best available solution to a problem subject to a certain set of constraints. For example, a linear programming model can be used to determine the optimal product mix in a production environment.

In many situations, two or more of the aforementioned models may be involved in the solution of a problem. For example, a descriptive model might provide insights into the nature of the problem, an optimization model might provide the optimal set of actions to take in solving the problem, a satisficing model might temper the optimal solution with reality, a prescriptive model might suggest the procedures for implementing the selected solution, and a predictive model might forecast a future outcome of the problem scenario.

STEP 5. MAKING THE DECISION

Using the available data, information, and the decision model, the decision-maker will determine the real-world actions that are needed to solve the stated problem. A sensitivity analysis may be useful for determining what changes in parameter values might cause a change in the decision.

STEP 6. IMPLEMENTING THE DECISION

A decision represents the selection of an alternative that satisfies the objective stated in the problem statement. A good decision is useless until it is implemented. An important aspect of a decision is to specify how it is to be implemented. Selling the decision and the project to management requires a well-organized persuasive presentation. The way a decision is presented can directly influence whether or not it is adopted. The presentation of a decision should include at least the following: an executive summary, technical aspects of the decision, managerial aspects of the decision, resources required to implement the decision, cost of the decision, the time frame for implementing the decision, and the risks associated with the decision.

Systems decisions are often complex, diffuse, distributed, and poorly understood. No one person has all the information to make all decisions accurately. As a result, crucial decisions are made by a group of people. Some organizations use outside consultants with appropriate expertise to make recommendations for important decisions. Other organizations set up their own internal consulting groups without having to go outside the organization. Decisions can be made through linear responsibility, in which case one person makes the final decision based on inputs from other people. Decisions can also be made through shared responsibility, in which case, a group of people share the responsibility for making joint decisions. The major advantages of group decision-making are listed as follows:

1. Facilitation of a systems view of the problem environment.
2. Ability to share experience, knowledge, and resources. Many heads are better than one. A group will possess greater collective ability to solve a given decision problem.
3. Increased credibility. Decisions made by a group of people often carry more weight in an organization.
4. Improved morale. Personnel morale can be positively influenced because many people have the opportunity to participate in the decision-making process.
5. Better rationalization. The opportunity to observe other people's views can lead to an improvement in an individual's reasoning process.
6. Ability to accumulate more knowledge and facts from diverse sources.
7. Access to broader perspectives spanning different problem scenarios.
8. Ability to generate and consider alternatives from different perspectives.
9. Possibility for a broader-base involvement, leading to a higher likelihood of support.

10. Possibility for group leverage for networking, communication, and political clout.

In spite of the much-desired advantages, group decision-making does possess the risk of flaws. Some possible disadvantages of group decision-making are listed as follows:

1. Difficulty in arriving at a decision
2. Slow operating timeframe
3. Possibility for individuals' conflicting views and objectives
4. Reluctance of some individuals in implementing the decision
5. Potential for power struggle and conflicts among the group
6. Loss of productive employee time
7. Too much compromise may lead to less-than-optimal group output
8. Risk of one individual dominating the group
9. Overreliance on group process may impede agility of management to make decision fast
10. Risk of dragging feet due to repeated and iterative group meetings

BRAINSTORMING

Brainstorming is a way of generating many new ideas. In brainstorming, the decision group comes together to discuss alternate ways of solving a problem. The members of the brainstorming group may be from different departments, may have different backgrounds and training, and may not even know one another. The diversity of the participants helps create a stimulating environment for generating different ideas from different viewpoints. The technique encourages free outward expression of new ideas no matter how far-fetched the ideas might appear. No criticism of any new idea is permitted during the brainstorming session. A major concern in brainstorming is that extroverts may take control of the discussions. For this reason, an experienced and respected individual should manage the brainstorming discussions. The group leader establishes the procedure for proposing ideas, keeps the discussions in line with the group's mission, discourages disruptive statements, and encourages the participation of all members.

After the group runs out of ideas, open discussions are held to weed out the unsuitable ones. It is expected that even the rejected ideas may stimulate the generation of other ideas, which may eventually lead to other favored ideas. Guidelines for improving brainstorming sessions are presented as follows:

Focus on a specific decision problem.
Keep ideas relevant to the intended decision.
Be receptive to all new ideas.
Evaluate the ideas on a relative basis after exhausting new ideas.
Maintain an atmosphere conducive to cooperative discussions.
Maintain a record of the ideas generated.

THE DELPHI METHOD

The traditional approach to group decision-making is to obtain the opinion of experienced participants through open discussions. An attempt is made to reach a consensus among the participants. However, open group discussions are often biased because of the influence of subtle intimidation from dominant individuals. Even when the threat of a dominant individual is not present, opinions may still be swayed by group pressure. This is called the "bandwagon effect" of group decision-making.

The Delphi method attempts to overcome these difficulties by requiring individuals to present their opinions anonymously through an intermediary. The method differs from the other interactive group methods because it eliminates face-to-face confrontations. It was originally developed for forecasting applications, but it has been modified in various ways for application to different types of decision-making. The method can be quite useful for project management decisions. It is particularly effective when decisions must be based on a broad set of factors. The Delphi method is normally implemented as follows:

1. *Problem definition*: A decision problem that is considered significant is identified and clearly described.
2. *Group selection*: An appropriate group of experts or experienced individuals is formed to address the particular decision problem. Both internal and external experts may be involved in the Delphi process. A leading individual is appointed to serve as the administrator of the decision process. The group may operate through the mail or gather together in a room. In either case, all opinions are expressed anonymously on paper. If the group meets in the same room, care should be taken to provide enough room so that each member does not have the feeling that someone may accidentally or deliberately observe their responses.
3. *Initial opinion poll*: The technique is initiated by describing the problem to be addressed in unambiguous terms. The group members are requested to submit a list of major areas of concern in their specialty areas as they relate to the decision problem.
4. *Questionnaire design and distribution*: Questionnaires are prepared to address the areas of concern related to the decision problem. The written responses to the questionnaires are collected and organized by the administrator. The administrator aggregates the responses in a statistical format. For example, the average, mode, and median of the responses may be computed. This analysis is distributed to the decision group. Each member can then see how his or her responses compare with the anonymous views of the other members.
5. *Iterative balloting*: Additional questionnaires based on the previous responses are passed to the members. The members submit their responses again. They may choose to alter or not to alter their previous responses.

6. *Silent discussions and consensus*: The iterative balloting may involve anonymous written discussions of why some responses are correct or incorrect. The process is continued until a consensus is reached. A consensus may be declared after five or six iterations of the balloting or when a specified percentage (e.g., 80%) of the group agrees on the questionnaires. If a consensus cannot be declared on a particular point, it may be displayed to the whole group with a note that it does not represent a consensus.

In addition to its use in technological forecasting, the Delphi method has been widely used in general decision-making. Its major characteristics of anonymity of responses, statistical summary of responses, and controlled procedure make it a reliable mechanism for obtaining numeric data from subjective opinion. The major limitations of the Delphi method are:

1. Its effectiveness may be limited in cultures where strict hierarchy, seniority, and age influence decision-making processes.
2. Some experts may not readily accept the contribution of nonexperts to the group decision-making process.
3. Since opinions are expressed anonymously, some members may take the liberty of making ludicrous statements. However, if the group composition is carefully reviewed, this problem may be avoided.

NOMINAL GROUP TECHNIQUE

The nominal group technique is a silent version of brainstorming. It is a method of reaching consensus. Rather than asking people to state their ideas aloud, the team leader asks each member to jot down a minimum number of ideas, for example, five or six. A single list of ideas is then written on a chalkboard for the whole group to see. The group then discusses the ideas and weeds out some iteratively until a final decision is made. The nominal group technique is easier to control. Unlike brainstorming where members may get into shouting matches, the nominal group technique permits members to silently present their views. In addition, it allows introversive members to contribute to the decision without the pressure of having to speak out too often.

In all of the group decision-making techniques, an important aspect that can enhance and expedite the decision-making process is the requirement that members review all pertinent data before coming to the group meeting. This will ensure that the decision process is not impeded by trivial preliminary discussions. Some disadvantages of group decision-making are:

1. Peer pressure in a group situation may influence a member's opinion or discussions.
2. In a large group, some members may not get to participate effectively in the discussions.
3. A member's relative reputation in the group may influence how well his or her opinion is rated.

4. A member with a dominant personality may overwhelm the other members in the discussions.
5. The limited time available to the group may create a time pressure that forces some members to present their opinions without fully evaluating the ramifications of the available data.
6. It is often difficult to get all members of a decision group together at the same time.

Despite the noted disadvantages, group decision-making definitely has many advantages that may nullify the shortcomings. The advantages as presented earlier will have varying levels of effect from one organization to another. Team work can be enhanced in group decision-making by adhering to the following guidelines:

1. Get a willing group of people together.
2. Set an achievable goal for the group.
3. Determine the limitations of the group.
4. Develop a set of guiding rules for the group.
5. Create an atmosphere conducive to group synergism.
6. Identify the questions to be addressed in advance.
7. Plan to address only one topic per meeting.

For major decisions and long-term group activities, arrange for team training that allows the group to learn the decision rules and responsibilities together. The steps for the nominal group technique are:

1. Silently generate ideas, in writing.
2. Record ideas without discussion.
3. Conduct group discussion for clarification of meaning, not argument.
4. Vote to establish the priority or rank of each item.
5. Discuss vote.
6. Cast final vote.

INTERVIEWS, SURVEYS, AND QUESTIONNAIRES

Interviews, surveys, and questionnaires are important information gathering techniques. They also foster cooperative working relationships. They encourage direct participation and inputs into project decision-making processes. They provide an opportunity for employees at the lower levels of an organization to contribute ideas and inputs for decision-making. The greater the number of people involved in the interviews, surveys, and questionnaires, the more valid the final decision. The following guidelines are useful for conducting interviews, surveys, and questionnaires to collect data and information for project decisions:

1. Collect and organize background information and supporting documents on the items to be covered by the interview, survey, or questionnaire.
2. Outline the items to be covered and list the major questions to be asked.

3. Use a suitable medium of interaction and communication: telephone, fax, electronic mail, face-to-face observation, meeting venue, poster, or memo.
4. Tell the respondent the purpose of the interview, survey, or questionnaire, and indicate how long it will take.
5. Use open-ended questions that stimulate ideas from the respondents.
6. Minimize the use of yes or no type of questions.
7. Encourage expressive statements that indicate the respondent's views.
8. Use the who, what, where, when, why, and how approach to elicit specific information.
9. Thank the respondents for their participation.
10. Let the respondents know the outcome of the exercise.

MULTIVOTE

Multivoting is a series of votes used to arrive at a group decision. It can be used to assign priorities to a list of items. It can be used at team meetings after a brainstorming session has generated a long list of items. Multivoting helps reduce such long lists to a few items, usually three to five. The steps for multivoting are:

1. Take a first vote. Each person votes as many times as desired but only once per item.
2. Circle the items receiving a relatively higher number of votes (i.e., majority vote) than the other items.
3. Take a second vote. Each person votes for a number of items equal to one-half the total number of items circled in Step 2. Only one vote per item is permitted.
4. Repeat Steps 2 and 3 until the list is reduced to three to five items depending on the needs of the group. It is not recommended to multivote down to only one item.
5. Perform further analysis of the items selected in Step 4, if needed.

CONCLUSION

In terms of a summary of a project formulation of the supply chain, systems integration is the synergistic linking together of the various components, elements, and subsystems of a system, where the system may be a complex project, a large endeavor, or an expansive organization. Activities that are resident within the system must be managed both from the technical and managerial standpoints. Any weak link in the system, no matter how small, can be the reason that the overall system fails. In this regard, every component of a project is a critical element that must be nurtured and controlled. Embracing the systems principles for project management will increase the likelihood of success of projects.

REFERENCES

Askin, Ronald G. and Jorge A. Sefair (2021), "Improving the efficiency of airport security screening checkpoints," *ISE Magazine*, Vol. 53, No. 5, pp. 26–31.

Badiru, Adedeji B. (1988), *Project Management in Manufacturing and High Technology Operations*, John Wiley & Sons, New York.

Badiru, Adedeji B. (2019a), *Project Management: Systems, Principles, and Applications*, 2nd Edition, Taylor & Francis Group/CRC Press, Boca Raton, FL.

Badiru, Adedeji B. (2019b), *Systems Engineering Models: Theory, Methods, and Applications*, Taylor & Francis Group/CRC Press, Boca Raton, FL.

NOAA – National Oceanic and Atmospheric Administration (2021), *Fishwatch: U.S. Seafood Facts*: https://www.fishwatch.gov/sustainable-seafood/the-global-picture, accessed 26 June 2021.

Zomorodi, Manoush and Fiona Geiran (2021), *Ayana Elizabeth Johnson: What Should You Look For When Shopping For Seafood?*, NPR Radio Hour podcast: https://www.wvxu.org/post/ayana-elizabeth-johnson-what-should-you-look-when-shopping-seafood#stream/0, accessed 26 June 2021.

2 Industrial Engineering Techniques in the Supply Chain

INTRODUCTION

Because of the diverse, flexible, and comprehensive systems-based methodologies of industrial engineering (IE), the discipline offers the best views and approaches for dealing with the challenges faced in the supply chain whether the challenges are due to unexpected disruptions, planned shutdowns, accidental occurrences, scheduling mismanagement, or deliberate sabotage. Each catastrophic event can often be addressed by the multiple and integrated methodologies of IE. For decades, IE has had a practical relationship with what is now popularly known as supply chain and logistics. In the early days, what we call logistics now was called "physical distribution" by industrial engineers. Consumers often erroneously think of the supply chain as being represented by the transportation and distribution aspects of providing products, services, and desired results. In fact, the supply chain involves both the upstream and downstream components. In this regard, the production end (the source) of the spectrum is as important as the transportation and distribution components (the destination).

What makes IE very applicable to the variety of supply chain problems is the fact that the discipline directly confronts who supplies whom with what, when, where, how, and why. The approach covers the spectrum of what can happen and proactively leverages contingency planning. The adaptive definition of IE presented below bears out this claim.

> Industrial Engineer – one who is concerned with the design, installation, and improvement of integrated systems of people, materials, information, equipment, and energy by drawing upon specialized knowledge and skills in the mathematical, physical, and social sciences, together with the principles and methods of engineering analysis and design to specify, predict, and evaluate the results to be obtained from such systems.

The above definition embodies the various aspects of what an industrial engineer does. For several decades, the military had called upon the discipline of IE to achieve program effectiveness and operational efficiencies. IE is versatile, flexible, adaptive, and diverse. It can be seen from the definition that a systems orientation permeates the work of industrial engineers. This is particularly of interest to the military because military operations and functions are constituted by linking systems. The major functions of industrial engineers include the following:

- Designing integrated systems of people, technology, process, and methods.
- Modeling operations to optimize cost, quality, and performance trade-offs.
- Developing performance modeling, measurement, and evaluation for systems.
- Developing and maintaining quality standards for government, industry, and business.
- Applying production principles to pursue improvements in service organizations.
- Incorporating technology effectively into work processes.
- Developing cost mitigation, avoidance, or containment strategies.
- Improving overall productivity of integrated systems of people, tools, and processes.
- Planning, organizing, scheduling, and controlling programs and projects.
- Organizing teams to improve efficiency and effectiveness of an organization.
- Installing technology to facilitate work flow.
- Enhancing information flow to facilitate smooth operation of systems.
- Coordinating materials and equipment for effective systems performance.
- Designing work systems to eliminate waste and reduce variability.
- Designing jobs (determining the most economic way to perform work).
- Setting performance standards and benchmarks for quality, quantity, and cost.
- Designing and installing facilities to incorporate human factors and ergonomics.

IE can be described as the practical application of the combination of engineering fields together with the principles of scientific management. It is the engineering of work processes and the application of engineering methods, practices, and knowledge to production and service enterprises. IE places a strong emphasis on an understanding of workers and their needs in order to increase and improve production and service activities. Figure 2.1 illustrates how IE serves as an umbrella discipline for many sub-specialties that apply directly to any complex supply chain. Some of the sub-areas are independent disciplines in their own rights. So coveted are the diverse skills of industrial engineers that they are sought after as process improvement engineers in hospitals, retail organizations, banking, and general financial services sectors.

IE has a proud heritage with a link that can be traced back to the *Industrial Revolution*. Although the practice of IE has been in existence for centuries, the

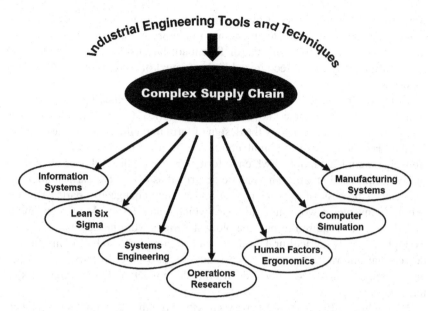

FIGURE 2.1 Industrial Engineering and Sub-specialties Applied to Complex Supply Chains.

work of Frederick Taylor in the early twentieth century was the first formal emergence of the profession. It has been referred to with different names and connotations. Scientific management was one of the early names used to describe what industrial engineers did.

Industry, the root of the profession's name, clearly explains what the profession is about. The dictionary defines industry generally as the ability to produce and deliver goods and services. The "industry" in IE can be viewed as the application of skills and creativity to achieve work objectives. This relates to how human effort is harnessed innovatively to carry out work. Thus, any activity can be defined as "industry" because it generates a product; be it service or physical product. A systems view of IE encompasses all the details and aspects necessary for applying skills and cleverness to produce work efficiently. The academic curriculum of IE continues to change, evolve, and adapt to the changing operating environment of the profession.

It is widely recognized that the occupational discipline that has contributed the most to the development of modern society is *engineering*, through its various segments of focus. Engineers design and build infrastructures that sustain the society. These include roads, residential and commercial buildings, bridges, canals, tunnels, communication systems, healthcare facilities, schools, habitats, transportation systems, and factories. Across all of these, the IE process of systems integration facilitates the success of the efforts. In this sense, the scope of IE goes through the levels of activity, task, job, project, program, process, system, enterprise, and society. This handbook of military IE presents essential tools for the levels embodied by this hierarchy of functions. From the age of horse-drawn carriages and steam engines to the present age of intelligent automobiles and aircraft, the impacts of IE cannot be mistaken, even though the contributions may not be recognized in the context of the conventional work of industry.

Going further back in history, several developments helped form the foundation for what later became known as IE. In America, George Washington was said to have been fascinated by the design of farm implements on his farm in Mt. Vernon. He had an English manufacturer send him a plow built to his specifications that included a mold on which to form new irons when old ones were worn out or would need repairs. This can be described as one of the earliest attempts to create a process of achieving a system of interchangeable parts. Thomas Jefferson invented a wooden mold board which, when fastened to a plow, minimized the force required to pull the plow at various working depths. This is an example of early agricultural industry innovation. Jefferson also invented a device that allowed a farmer to seed four rows at a time. In pursuit of higher productivity, he invented a horse-drawn threshing machine that did the work of ten men.

Meanwhile in Europe, productivity growth, through reductions in manpower, marked the technological innovations of the 1769–1800 Europe. Sir Richard Arkwright developed a practical code of factory discipline. In their foundary, Matthew Boulton and James Watt, developed a complete and integrated engineering plant to manufacture steam engines. They developed extensive methods of market research, forecasting, plant location planning, machine layout, work flow, machine operating standards, standardization of product components, worker training, division of labor, work study, and other creative approaches to increasing productivity. Charles Babbage, who is credited with the first idea of a computer, documented ideas on scientific methods of managing industry in his book entitled On the Economy of Machinery and Manufacturers, which was first published in 1832. The book contained ideas on division of labor, paying less for less important tasks, organization charts, and labor relations. These were all forerunners of modern IE.

In the early history of the United States, several efforts emerged to form the future of the IE profession. Eli Whitney used mass production techniques to produce muskets for the US Army. In 1798, Whitney developed the idea of having machines make each musket part so that it could be interchangeable with other similar parts. By 1850, the principle of interchangeable parts was widely adopted. It eventually became the basis for modern mass production for assembly lines. It is believed that Eli Whitney's principle of interchangeable parts contributed significantly to the Union victory during the US Civil War. Thus, the early practice of IE made significant contribution to the military. That heritage has continued until today.

The management attempts to improve productivity prior to 1880 did not consider the human element as an intrinsic factor. However, from 1880 through the first quarter of the twentieth century, the works of Frederick W. Taylor, Frank and

Lillian Gilbreth, and Henry L. Gantt created a long-lasting impact on productivity growth through consideration of the worker and his or her environment.

Frederick Winslow Taylor (1856–1915) was born in the Germantown section of Philadelphia to a well-to-do family. At the age of 18, he entered the labor force, having abandoned his admission to Harvard University due to an impaired vision. He became an apprentice machinist and pattern-maker in a local machine shop. In 1878, when he was22, he went to work at the Midvale Steel Works. The economy was in a depressed state at the time. Frederick was employed as a laborer. His superior intellect was very quickly recognized. He was soon advanced to the positions of time clerk, journeyman, lathe operator, gang boss, and foreman of the machine shop. By the age of 31, he was made the chief engineer of the company. He attended night school and earned a degree in mechanical engineering in 1883 from Stevens Institute. As a work leader, Taylor faced the following common questions:

1. "Which is the best way to do this job?"
2. "What should constitute a day's work?"

These are still questions faced by industrial engineers of today. Taylor set about the task of finding the proper method for doing a given piece of work, instructing the worker in following the method, maintaining standard conditions surrounding the work so that the task could be properly accomplished, and setting a definite time standard and payment of extra wages for doing the task as specified. Taylor later documented his industry management techniques in his book entitled *The Principles of Scientific Management*.

The work of Frank and Lillian Gilbreth coincided with the work of Frederick Taylor. In 1895, on his first day on the job as a bricklayer, Frank Gilbreth noticed that the worker assigned to teach him how to lay brick did his work in three different ways. The bricklayer was insulted when Frank tried to tell him of his work inconsistencies – when training someone on the job, when performing the job himself, and when speeding up. Frank thought it was essential to find one best way to do work. Many of Frank Gilbreth's ideas were similar to Taylor's ideas. However, Gilbreth outlined procedures for analyzing each step of work flow. Gilbreth made it possible to apply science more precisely in the analysis and design of the work place. Developing *therbligs*, which is a moniker for Gilbreth spelled backward, as elemental predetermined time units, Frank and Lillian Gilbreth were able to analyze the motions of a worker in performing most factory operations in a maximum of 18 steps. Working as a team, they developed techniques that later became known as work design, methods improvement, work simplification, value engineering, and optimization. Lillian (1878–1972) brought to the engineering profession the concern for human relations. The foundation for establishing the profession of IE was originated by Frederick Taylor and Frank and Lillian Gilbreth.

The work of Henry Gantt (1861–1919) advanced the management movement from an industrial management perspective. He expanded the scope of managing industrial operations. His concepts emphasized the unique needs of the worker by recommending the following considerations for managing work:

1. Define his task, after a careful study.
2. Teach him how to do it.
3. Provide an incentive in terms of adequate pay or reduced hours.
4. Provide an incentive to surpass it.

Henry Gantt's major contribution is the Gantt Chart, which went beyond the works of Frederick Taylor or the Gilbreths. The Gantt chart related every activity in the plant to the factor of time. This was a revolutionary concept for the time. It led to better production planning control and better production control. This involved visualizing the plant as a whole, like one big system made up of inter-related sub-systems. The major chronological historical, scholarly, intellectual, and practical developments marking the applications of IE are summarized below, albeit not necessarily under the name of "industrial engineering." The essence of the profession is what matters as highlighted in the chronology below, which contains several elements of building, maintaining, and sustaining supply chains in business and industry:

1440:	Venetian ships are reconditioned and refitted on an assembly line.
1474:	Venetian Senate passes the first patent law and other industrial laws.
1568:	Jacques Besson publishes illustrated book on iron machinery as replacement for wooden machines.
1722:	Rene de Reaunur publishes the first handbook on iron technology.
1733:	John Kay patents the flying shuttle for textile manufacture – a landmark in textile mass production.
1747:	Jean Rodolphe Perronet establishes the first engineering school.
1765:	Watt invents the separate condenser, which made the steam engine the power source.
1770:	James Hargreaves patents his "Spinning Jenny." Jesse Ramsden devises a practical screw-cutting lathe.
1774:	John Wilkinson builds the first horizontal boring machine.
1775:	Richard Arkwright patents a mechanized mill in which raw cotton is worked into thread.
1776:	James Watt builds the first successful steam engine, which became a practical power source; Adam Smith discusses the division of labor in *The Wealth of Nations*.
1785:	Edmund Cartwright patents a power loom.
1793:	Eli Whitney invents the "cotton gin" to separate cotton from its seeds.
1797:	Robert Owen uses modern labor and personnel management techniques in a spinning plant in the New Lanark Mills in Manchester, England.
1798:	Eli Whitney designs muskets with interchangeable parts.
1801:	Joseph Marie Jacquard designs automatic control for pattern-weaving looms using punched cards.
1802:	"Health and Morals Apprentices Act" in Britain aims at improving standards for young factory workers; Marc Isambard Brunel, Samuel Benton, and Henry Maudsey design an integrated series of 43 machines to mass produce pulley blocks for ships.

1818: Institution of Civil Engineers founded in Britain.
1824: The repeal of the Combination Act in Britain legalizes trade unions.
1829: Mathematician Charles Babbage designs "analytical engine," a forerunner of the modern digital computer.
1831: Charles Babbage publishes *On the Economy of Machines and Manufacturers*.
1832: The Sadler Report exposes the exploitation of workers and the brutality practiced within factories.
1833: Factory law enacted in the United Kingdom. The Factory Act regulates British children's working hours; a general Trades Union is formed in New York.
1835: Andrew Ure publishes "The Philosophy of Manufacturers," described the new industrial system that had developed in England over the course of the previous century. He advised that the new factory system is beneficial to workers because it relieved them of much of the tedium of manufacturing goods by hand. It is in this same year that Samuel Morse invented the telegraph, which opened up communication between remote locations. This invention is the forerunner of digital communication as we know it today.
1845: Friederich Engels publishes *Condition of the Working Classes in England*, which represented a rigorous formal a study of the industrial working class in Victorian England. This same sense of paying attention to the conditions, plight, and opportunities of workers can still be seen in the IE of today.
1847: The British Government passed the Factory Act to improve conditions of women and children working in factories. Young children were working very long hours in deplorable working environments. Under the Act, no child workers under 9 years of age were allowed and employers must have an age certificate for their child workers. Children of 9–13 years to work no more than 9 hours a day; George Stephenson founded the Institution of Mechanical Engineers. He was a British civil engineer and mechanical engineer. Renowned as the "Father of Railways," Stephenson was considered by the Victorians a great example of diligent application and thirst for improvement. It can be seen that the early works of IE mirror the early industrial focus of mechanical engineering. The "thirst for improvement" is still seen in today's IE commitment to continuous improvement.
1948: Frank B. Gilbreth, Jr. and Ernestine Gilbreth Carey publish *Cheaper by the Dozen*, which celebrates the professional and family accomplishments of their parents, Frank and Lillian Gilbreth.
1856: Henry Bessemer revolutionizes the steel industry through a novel design for a converter.
1869: Transcontinental railroad completed in the United States.
1871: British Trade Unions are legalized by Act of Parliament.
1876: Alexander Graham Bell invents a usable telephone. As an advancement over Samuel Morse's telegraph, the usable telephone facilitated and expedited communication beyond comprehension in that era.
1877: Thomas Edison invents the phonograph. The phonograph was developed as a result of Thomas Edison's work on two other inventions, the telegraph and the telephone. In 1877, Edison was working on a machine that would transcribe telegraphic messages through indentations on paper tape, which could later be sent over the telegraph repeatedly. The recorded voice quickly became a tool for operational improvements in industrial establishments.

1878:   Frederick W. Taylor joins Midvale Steel Company. Taylor was widely known for
        his methods to improve industrial efficiency. He was one of the first
        management consultants. Taylor was one of the intellectual leaders of the
        Efficiency Movement and his ideas, broadly conceived, were highly influential
        in the Progressive Era of the 1890s to the 1920s. Efficiency, in any form,
        whether digital or analog, has remained a focus in the practice of IE of today.
1880:   American Society of Mechanical Engineers (ASME) is organized.
1881:   Frederick Taylor begins time study experiments that would lay the foundation
        for the early introduction of IE to industries.
1885:   Frank B. Gilbreth begins motion study research. A **time and motion study** (or
        time-motion study) is an efficiency technique that combined the Time Study
        work of Frederick Taylor with the Motion Study work of Frank and Lillian
        Gilbreth. It was a major part of Scientific Management (also known as
        Taylorism). After its first introduction, time study developed in the direction of
        establishing standard times for work elements, while motion study evolved into
        a technique for improving work methods. The two techniques became
        integrated and refined into a widely accepted method applicable to the
        improvement and upgrading of work systems, which is still practiced by IE
        today under the name of methods engineering, which can be found in industrial
        establishments, service organizations, banks, schools, hospitals, government,
        and the military. The tools and techniques of IE are everywhere.
1886:   Henry R. Towne presents the paper, *The Engineer as Economist*; the American
        Federation of Labor (AFL) is organized; Vilfredo Pareto publishes *Course in
        Political Economy*; Charles M. Hall and Paul L. Herault independently invent
        an inexpensive method of making aluminum.
1888:   Nikola Tesla invents the alternating current induction motor, enabling electricity
        to take over from steam as the main provider of power for industrial machines;
        Dr. Herman Hollerith invents the electric tabulator machine, the first successful
        data processing machine.
1890:   Sherman Anti-Trust Act is enacted in the United States.
1892:   Gilbreth completes motion study of bricklaying.
1893:   Taylor begins work as a consulting engineer.
1895:   Taylor presents paper entitled *A Piece-Rate System* to ASME.
1898:   Taylor begins time study at Bethlehem Steel; Taylor and Maunsel White develop
        process for heat-treating high-speed tool steels.
1899:   Carl G. Barth invents a slide rule for calculating metal cutting speed as part of
        Taylor system of management.
1901:   American national standards are established; Yawata Steel begins operation in
        Japan.
1903:   Taylor presents paper entitled *Shop Management* to ASME; H. L. Gantt
        develops the "Gantt Chart"; Hugo Diemers writes *Factory Organization and
        Administration*; Ford Motor Company is established.
1904:   Harrington Emerson implements Santa Fe Railroad improvement; Thorstein B.
        Veblen: *The Theory of Business Enterprise*.
1906:   Taylor establishes metal-cutting theory for machine tools; Vilfredo Pareto
        publishes *Manual of Political Economy*.
1907:   Frank Gilbreth uses time study for construction.
1908:   Model T Ford is built; Pennsylvania State College introduces the first university
        course in IE.
1911:   Frederick Taylor publishes *The Principles of Scientific Management*; Frank
        Gilbreth publishes *Motion Study*; Factory laws are enacted in Japan.

1912: Harrington Emerson publishes *The Twelve Principles of Efficiency*; Frank and Lillian Gilbreth presented the concept of "therbligs"; Yokokawa translates into Japanese Taylor's *Shop Management* and *The Principles of Scientific Management*.

1913: Henry Ford establishes a plant at Highland Park, Michigan, which utilizes the principles of uniformity and interchangeability of parts, and of the moving assembly line by means of conveyor belt; Hugo Munstenberg publishes *Psychology of Industrial Efficiency*.

1914: World War I starts; Clarence B. Thompson edits *Scientific Management*, a collection of articles on Taylor's system of management.

1915: Taylor's system is used at Niigata Engineering's Kamata plant in Japan; Robert Hoxie publishes *Scientific Management and Labour*; Lillian Gilbreth earned Ph.D. at Brown U in 1915 in psychology.

1916: Lillian Gilbreth publishes *The Psychology of Management*; Taylor Society established in the United States.

1917: Frank and Lillian Gilbreth publish *Applied Motion Study*; The Society of Industrial Engineers is formed in the United States of America.

1918: Mary P. Follet publishes *The New State: Group Organization, the Solution of Popular Government*.

1919: Henry L. Gantt publishes *Organization for Work*.

1920: Merrick Hathaway presents paper: *Time Study as a Basis for Rate Setting*; General Electric establishes divisional organization; Karel Capek introduces *Rossum's Universal Robots*. This play coined the word "robot."

1921: The Gilbreths introduce process-analysis symbols to ASME.

1922: Toyoda Sakichi's automatic loom is developed; Henry Ford publishes *My Life and Work*.

1924: Frank and Lillian Gilbreth announce results of micromotion study using therbligs; Elton Mayo conducts illumination experiments at Western Electric.

1926: Henry Ford publishes *Today and Tomorrow*.

1927: Elton Mayo and others begin relay-assembly test room study at the Hawthorne plant.

1929: Great Depression; International Scientific Management Conference held in France.

1930: Hathaway publishes *Machining and Standard Times*; Allan H. Mogensen discusses 11 principles for work simplification in *Work Simplification*; Henry Ford publishes *Moving Forward*

1931: Dr. Walter Shewhart publishes *Economic Control of the Quality of Manufactured Product*.

1932: Aldous Huxley publishes *Brave New World*, the satire which prophesies a horrifying future ruled by industry.

1934: General Electric performs micromotion studies.

1936: The word "automation" is first used by D. S. Harder of General Motors. It is used to signify the use of transfer machines which carry parts automatically from one machine to the next, thereby linking the tools into an integrated production line; Charlie Chaplin produces *Modern Times*, a film showing an assembly line worker driven insane by routine and unrelenting pressure of his job.

1937: Ralph M. Barnes publishes *Motion and Time Study*.

1941: R. L. Morrow publishes *Ratio Delay Study*, an article in *Mechanical Engineering* journal; Fritz J. Roethlisberger: *Management and Morale*.

1943: ASME work standardization committee publishes glossary of IE terms.

1945:   Marvin E. Mundel devises "memo-motion" study, a form of work measurement
        using time-lapse photography; Joseph H. Quick devises work factors (WF)
        method; Shigeo Shingo presents concept of production as a network of
        processes and operations and identifies lot delays as source of delay between
        processes, at a technical meeting of the Japan Management Association.
1946:   The first all-electronic digital computer ENIAC (Electronic Numerical Integrator
        and Computer) is built at Pennsylvania University; the first fully automatic
        system of assembly is applied at the Ford Motor Plant.
1947:   American mathematician, Norbert Wiener publishes *Cybernetics*.
1948:   H. B. Maynard and others introduce methods time measurement (MTM)
        method; Larry T. Miles develops value analysis (VA) at General Electric;
        Shigeo Shingo announces process-based machine layout; the American
        Institute of Industrial Engineers is formed.
1950:   Marvin E. Mundel publishes *Motion and Time Study, Improving Productivity*.
1951:   Inductive statistical quality control is introduced to Japan from the
        United States
1952:   Role and sampling study of IE conducted at ASME.
1953:   B. F. Skinner: *Science of Human Behavior*.
1956:   New definition of IE is presented at the American Institute of Industrial
        Engineers Convention: "Industrial engineering is concerned with the design,
        improvement and installation of integrated systems of people, material,
        information, equipment and energy. It draws upon specialized knowledge and
        skills in the mathematical, physical and social sciences, together with the
        principles and methods of engineering analysis and design to specify, predict
        and evaluate the results to be obtained from such systems."
1957:   Chris Argyris: *Personality and Organization*; Herbert A. Simon: *Organizations*;
        R. L. Morrow: *Motion and Time Study*; Shigeo Shingo introduces scientific
        thinking mechanism (STM) for improvements; The Treaty of Rome established
        the European Economic Community.
1960:   Douglas M. McGregor publishes *The Human Side of Enterprise*.
1961:   Rensis Lickert publishes *New Patterns of Management*; Shigeo Shingo devises
        ZQC (source inspection and poka-yoke systems; Texas Instruments patents the
        silicon chip integrated circuit).
1963:   H. B. Maynard publishes *Industrial Engineering Handbook*; Gerald Nadler
        publishes *Work Design*.
1964:   Abraham Maslow publishes *Motivation and Personality*.
1965:   Transistors are fitted into miniaturized "integrated circuits."
1966:   Frederick Hertzberg publishes *Work and the Nature of Man*.
1967:   Sidney Gordon Gilbreath III completes Ph.D. dissertation on *A Bayesian
        Procedure for the Design of Sequential Sampling Plans* at Georgia Institute of
        Technology.
1968:   Roethlisberger: *Man in Organizations*; Department of Defense publishes
        Principles and Applications of Value Engineering.
1969:   Shigeo Shingo develops single-minute exchange of dies (SMED); Shigeo
        Shingo introduces pre-automation; Wickham Skinner publishes
        Manufacturing—missing link in corporate strategy article in *Harvard Business
        Review*.
1971:   Taiichi Ohno completes the Toyota production system; Intel Corporation
        develops the microprocessor chip.
1973:   First annual Systems Engineering Conference of the AIIE.

1975: Shigeo Shingo extols NSP-SS (non-stock production) system; Joseph Orlicky publishes *MRP: Material Requirements Planning*.

1976: IBM markets the first Personal Computer.

1980: Matsushita Electric used Mikuni method for washing machine production; Shigeo Shingo: *Study of the Toyota Production System from an Industrial Engineering Viewpoint*.

1981: Oliver Wight publishes *Manufacturing Resource Planning (MRP II)*; the AIIE (American Institute of Industrial Engineers) became the IIE (Institute of Industrial Engineers).

1982: Gavriel Salvendy publishes *Handbook of Industrial Engineering*.

1984: Shigeo Shingo: *A Revolution in Manufacturing* publishes *The SMED System*; Emergence of more formal practice area named Hospital Industrial Engineering to leverage IE tools and techniques for operational improvement in the healthcare industry. National hospital's operational excellence has since been credited to the application of IE.

1989: Development of Code Division Multiple Access (CDMA) for Cellular Communications.

1990: Wide use of the concept of Total Quality Management (TQM); Kjell Zandin publishes *MOST Work Measurement Systems: Basic Most, Mini Most, Maxi Most*.

1993: Adedeji Badiru and B. J. Ayeni publish *Practitioner's Guide to Quality and Process Improvement*.

1995: Adedeji Badiru publishes *Industry's Guide to ISO* 9000; The dot-com boom started in earnest; Netscape search engine was introduced; Peter Norvig and Stuart Norvig publishes *Artificial Intelligence: A Modern Approach*, which later became the authoritative textbook on AI.

2000: The turning point of the twenty-first century and the Y2K computer date scare.

2004: The birth of Facebook social networking; Skype took over worldwide online communication.

2008: The National Academy of Engineering (NAE) publishes the 14 Grand Challenges for Engineering; Adedeji Badiru et al. publish *Industrial Project Management: Concepts, Tools, and Techniques*.

2009: Adedeji Badiru and Marlin Thomas publishes the *Handbook of Military Industrial Engineering* to promote the application of IE in national defense strategies, which won the 2010 Book Of The Year Award from the IISE.

2012: Adedeji Badiru et al. publish *Industrial Control Systems: Mathematical and Statistical Models and Techniques*.

2014: Adedeji Badiru publishes the second edition of the *Handbook of Industrial and Systems Engineering*; Unmanned Aerial Vehicles (UAVs aka Drones) emerged as practical for a variety of applications.

2016: The IIE (Institute of Industrial Engineers) changed name to the IISE (Institute of Industrial & Systems Engineering); Wide appearance of self-driving cars; Adedeji Badiru publishes *Global Manufacturing Technology Transfer*.

2017: Adedeji Badiru and Sharon Bommer publish *Work Design: A Systematic Approach*; A. Ravi Ravindran and Donald Warsing publish *Supply Chain Engineering: Models and Applications*; Internet of Things (IoT) makes a big splash.

2018: Emergence of hybrid academic programs encompassing IE, digital engineering, data analytics, and virtual reality simulation.

2019:	Adedeji Badiru publishes *The Story of Industrial Engineering*, which won the 2020 Book Of The Year award from the IISE; Adedeji Badiru and Cassie Barlow edit *Defense Innovation Handbook: Guidelines, Strategies, and Techniques*; Adedeji Badiru publishes *Systems Engineering Models: Theory, Methods, and Applications*; Adedeji Badiru et al. publish *Manufacturing and Enterprise: An Integrated Systems Approach*.
2020:	Tools and techniques of IE applied to supply chain networks related to healthcare emergency needs necessitated by the COVID-19 pandemic; Adedeji Badiru publishes *Innovation: A Systems Approach*.
2021:	Adedeji Badiru and Lauralee Cromarty publish *Operational Excellence in the new Digital Era*; Adedeji Badiru and Tina Agustiady publish *Sustainability: A Systems Engineering Approach to the Global Grand Challenge*; Adedeji Badiru publishes *Data Analytics: Handbook of Formulas and Techniques*.
2022:	Adedeji Badiru publishes *Global Supply Chain: Using Systems Engineering Strategies to respond to Disruptions*.

As can be seen from the historical details above, IE has undergone progressive transformation over the past several decades and its applicability to the supply chain system is not in question. Frank and Lillian Gilbreth, whose names appear several times in the chronological accounts, were industrial engineers. They were among the first in the scientific field of operations management and the first in motion study and analysis. From 1910 to 1924, their company, Gilbreth, Inc., was employed as "efficiency experts" by many of the major industrial plants in the United States, Britain, and Germany. When Frank Gilbreth died in 1924, his wife, Lillian, carried on the work and became, perhaps, the foremost woman industrial engineer. She remained professionally active until her death in 1972.

PRODUCTION, QUALITY INSPECTION, AND THE SUPPLY CHAIN

Any supply chain relies on the production of products that the market wants. The shipment of finished goods is under the assumption that the products destined for shipment (through the supply chain) are of acceptable quality. Rejected quality ultimately affects the ability of the producer to meet market demands for the products with a reasonable time span. As long ago as 1976, S. G. Gilbreath (Gilbreath, 1967) recognized this linkage and did an extensive doctoral research on acceptance sampling. Although industry has been moving away from acceptance sampling under the pretext of focusing on using lean-six-sigma techniques to produce products of acceptable quality, Gilbreath's foundational research still offers some seminal insights into the classical role of quality inspection at the source. With good inspection practices, more acceptable products can be generated to keep the supply chain humming well, even when there are unexpected disruptions, such as the COVID-19 pandemic. An excerpt of Gilbreath's research results is presented here as a guide, hopefully, for future researchers interested in linking quality inspection to the robustness of modern supply chains. This is a gem of ideas and leads relevant idea for contemporary research studies in the context of prevailing global supply chains, driven by modern interconnectedness of producers, markets, and shipping services.

In Gilbreath's research (Gilbreath, 1967 and references therein), mathematical models for item-by-item sequential sampling of attributes are developed. The models consider prior distributions of process quality, inspection costs, and decision losses. Models are presented in which the prior distributions are:

(1) Hypergeometric
(2) Binomial
(3) Polya
(4) mixed binomial

Bayesian decision rules are the bases for computing expected losses. A conceptual model for identifying and measuring inspection costs and decision losses is formulated. This model is applied to a hypothetical example (Gilbreath, 1967). The work of previous investigators of sampling procedures who have considered prior distributions and costs is extended in four ways. These are:

(1) Development of a concept for identifying and measuring inspection costs and decision losses.
(2) Design of general cost models for item-by-item sequential sampling of attributes considering prior distributions, inspection costs, and decision losses.
(3) Design of a computer program of the sampling models in order that they be readily adaptable to inspection procedures using online access data processing equipment.
(4) Analysis of the sensitivity of the sampling models to changes in decision losses.

The difficulty in identifying and measuring relevant cost criteria for sampling inspection decisions may, in many cases, be attributed to attempts by researchers to consider too broad a spectrum of possible applications. By first outlining a procedure for establishing a model of a specific enterprise's situation and then analyzing that particular model, the determination and allocation of economic criteria are rendered far less formidable. If relatively accurate criteria can be obtained, they are, in the opinion of many researchers of sampling procedures, superior to statistical criteria as the bases for designing or choosing sampling inspection procedures. The relevant economic criteria to be considered are:

(1) C_f, the fixed sampling cost,
(2) C_s, the variable sampling cost, per item inspected,
(3) C_a, the decision loss accompanying acceptance of a defective item, and
(4) C_r, the decision loss resulting from rejection of a good item.

The decision to be made at any point in the course of inspection is among the alternatives:

(1) accept the lot now
(2) reject the lot now
(3) inspect one more item and accept or reject the lot

The alternative decision carrying the lowest total expected cost, where total expected cost is the sum of inspection costs and decision losses is selected. If (1) or (2) is most economical, inspection is terminated. If (3) is most economical, one more item is inspected. Although (3) is defined for purposes of cost evaluation as "inspect one more item and accept or reject the lot," actual acceptance or rejection is deferred until (1), (2), and (3) are re-evaluated for each new sample. Inspection continues until at some point decision (1) or (2) is made and the lot is categorized appropriately. N is the lot size, n is the accumulative sample size, x is the cumulative number of defective items in the accumulated sample, y is the number of defectives in (N-n), the uninspected portion of a lot, and is the probability of acceptance or is the probability that the next, (n+1)st, item inspected is good given that x defective items have been found in a cumulative sample of n items. X is the number of defective items in a lot.

GILBREATH'S DECISION MODEL

For: $n = (0, 1, ..., N)$,
$x = [\text{Max.} (0, n - N + x), ... \text{Min.} (n, x)]$
$X = (0, 1, ..., N)$, and
$y = (X - x)$
the expected costs and losses of the three alternative decisions are:

(1) $C_a E(y|x_n)$, where the subscript n indicates the result corresponds to a cumulative sample of size n,
(2) $C_r[(N - n) - E(y|x_n)]$
(3) $C_f + Cs + Ca\, E[y|(x = 0)_{n=1}]P_A + c_r\{(N - 1) - E[y|(x = 1)_{n=1}]\}(1 - P_A)$, for $n = 0$

$$C_s + C_a E\left(y|X_{n+1}\right)P_A + C_r\left[N - n - 1\right) - E\left(y|x + l_{n+1}\right)](1 - P_A)\,\text{for } n = (1,2,...,N)$$

At any value of n when either (1) or (2) becomes cheaper than (3), inspection ceases and the lot is categorized appropriately. This iterative procedure is computer programmed to adapt the sequential plans to online access equipment. The sensitivity analysis consists of demonstrating the response of the proposed model to shifts in the relevant statistical and economic parameters. Several specific cases are examined.

The sequential procedure was compared to categorizing without inspection and to a system of optimum single sampling plans for the two-point mixed binomial prior distribution over a variety of statistical and economic parameters. The proposed plans were in no case less economical than categorizing without inspection. For all conditions tested, the proposed procedure was significantly inferior to the optimum single sampling plans. Sensitivity of the sequential model to changes in quality levels of submitted lots and to changes in decision losses was favorable under all conditions tested.

JUSTIFICATION FOR ACCEPTANCE SAMPLING

The purpose of Gilbreath's research (Gilbreath, 1967) is to contribute to industrial understanding of and ability to use item-by-item sequential sampling. This is accomplished by providing a model, whereby in practice, item-by-item sequential sampling plans can be designed when certain costs are known and a specific prior distribution of product quality exists. A conceptual model for determining relevant costs and losses is presented.

The objective of Gilbreath's study is to develop a mathematical model which will specify an item-by-item sequential sampling plan to be used in a given situation. Samples will be considered to be drawn from lots selected randomly from a process thereby generating a probability distribution of product quality. Hald's compound hypergeometric distribution (Gilbreath, 1967) describes the sampling procedure. The specific objectives of Gilbreath's research are:

(1) Development of a concept for allocating inspection costs and decision losses on a per-occurrence basis.
(2) Design of cost models for sequential sampling of attributes considering prior distributions, inspection costs, and decision losses.
(3) Comparison of the sequential plans developed herein with the optimum single plans presented by A. Hald (Gilbreath, 1967).
(4) Computer programming of the sequential sampling model to demonstrate the adaptability of the model to online access data processing equipment.
(5) Analysis of the sensitivity of the sequential sampling models to shifts in allocated decision losses.

Acceptance sampling in statistical quality control has been the subject of much effort and scientific inquiry since industrialists, quality control practitioners, and statisticians first became aware that, under certain circumstances, it is cheaper to inspect samples of product and "take the chance" that the sample is truly representative of the lot than to examine every item in the lot (Gilbreath, 1967 and all references therein). Many rules have been established for determining sample sizes and acceptance or rejection criteria. There are almost as many bases for such rules as there are rules themselves. There are plans that give a certain "assurance" of accepting lots, whose quality is better than some minimum desirable level. There are plans that give another certain "assurance" that lots worse than some maximum desirable quality level will be rejected. Other plans attempt to give both these assurances and, in most cases because the sampling distributions are discrete, result in some minor compromise in respect to at least one of these assurances.

There are many formulas for finding the expected costs of such sampling plans and in specific cases they may be helpful in determining whether it is cheaper to inspect 100%, to sample, or not to inspect at all. There is a serious limitation in most of these cost determinations, however. They depend upon some prior knowledge (or guess) concerning the true lot product quality. That is to say, such cost determinations are based upon conditional probability, the probability of lot acceptance given a specific lot quality.

In 1960, A. Hald, at the University of Copenhagen (per Gilbreath, 1967), based single sampling plans upon a model more closely resembling reality. Industrial processes tend to produce quality within some process capability. As long as this process capability is the result of a constant cause system, product quality will generate a probability distribution. The probability that there are X defectives in a lot = p(X). Furthermore, if there are N items in a lot, X may vary from 0 to N or, $0 < X < N$. When a sample of n items is taken from the lot, the probability that the sample contains x defective items = p(x|X) where $0 < x < n$ and $0 < x < X$. In other words, if only this last probability is considered, then some assumption must be made about X, the number of defectives in the lot. This is the approach taken by investigators prior to the work of Hald. By considering p(X), the prior distribution, Hald approached more closely the true sampling situation that the probability of x defectives in a sample equals the probability of x defectives in a sample given all possible values of X defectives in the lot from which the sample was drawn. Letting x = number of defectives in the sample and y = number of defectives in the remainder of the lot,

$$X = x + y$$

and

$$p(x,y) = PN(x+y)\begin{bmatrix} n \\ x \end{bmatrix}\begin{bmatrix} N-n \\ y \end{bmatrix} / \begin{bmatrix} N \\ x+y \end{bmatrix}$$

from which Hald derives the marginal distribution of x:

$$gN(x) = \begin{bmatrix} n \\ x \end{bmatrix}\sum_{y=0}^{N-n} PN(x+y)\begin{bmatrix} N-n \\ y \end{bmatrix} / \begin{bmatrix} N \\ x+y \end{bmatrix}$$

The costs and losses to be considered are those contributing to product cost and related to product quality. They are:

1. The loss incurred by accepting bad lots.
2. The loss incurred by rejecting good lots.
3. The costs of inspecting items.

An inspection procedure that minimizes an appropriate function of these costs is desired when it is appropriate to select an inspection procedure. If all criteria are reduced to the three economic ones above, a model or models to minimize the total cost function would be universally applicable. According to Gilbreath (1967), Hald has formulated such models for single sampling. At the conclusion of his work, he suggested that his models be extended so as to apply to sequential sampling. It is toward such an extension that the proposed research is directed.

If the objectives of this research are accomplished, the resulting sequential sampling model, together with Hald's models, will provide economic bases for selecting single or sequential sampling plans. This would constitute a large and practical step forward, toward reality, from existing sampling plan selection techniques based upon somewhat nebulous probabilistic considerations related to economics only through implication.

It is the intent of Gilbreath's research to extend the then-current practice in item-by-item sequential sampling to include consideration of prior distribution and costs. The cost ratios used by Hald in his single sampling work is investigated relative to their extended applicability to sequential analysis. The binomial, hypergeometric, Polya, and mixed binomial prior distributions investigated by Hald, Johnson, and Schafer (see Gilbreath, 1967) will be applied to sequential sampling practice. A Bayesian statistical model will be designed, whereby sequential sampling plans may be designed considering prior distribution and using cost as the measure of effectiveness.

Gilbreath's study considers item-by-item sequential sampling for the attributes, good or bad. Acceptance sampling by attributes has been studied carefully to develop an understanding of the strengths and weaknesses of then-existing procedures for designing acceptance sampling plans. Consideration is given design schemes which do not depend explicitly upon economic criteria and others which do. Careful attention is given to a study of the state of the art (science) of the identification and measurement of costs and losses comprising relevant criteria for acceptance of sampling decisions. Further investigations have been made into the usefulness of existing criteria and measurement techniques. The study is limited to product inspection where the inspection is performed for the purpose of accepting lots of products containing few defective items and rejecting those showing indications of containing many. No attention was given to variables inspection designs. No consideration is given to the psychological aspects of acceptance sampling. This may be a good lead for modern research focusing on human factors and ergonomics in industrial settings. Gilbreath's research is confined to the economic, statistical, and mathematical theories of acceptance sampling for attributes.

The inspection of product could occur in a wide variety of circumstances. The product might be incoming purchased material, outgoing product undergoing final inspection, or product at any stage within a manufacturing process. It is assumed that the process responsible for the product's characteristics up until it is inspected is stable to the degree that a process curve can exist. By process curve is meant f(p), the probability density function for the fraction defective, p.

ACCEPTANCE SAMPLING IN THE LITERATURE

Gilbreath (1967) presents a comprehensive survey of acceptance sampling. Acceptance sampling literature was reviewed and classified according to the content of each publication. The material studied was primarily classified according to the criteria employed by decision rules. These criteria are either statistical or economic. Statistical criteria employed in designing acceptance sampling plans are similar to those used in significance testing. The risks or probabilities of commission of certain errors are fixed, and sample sizes are determined along with acceptance and rejection criteria to achieve these risks.

SINGLE, DOUBLE, AND MULTIPLE SAMPLING

Acceptance sampling by attributes has been discussed extensively. However, most of what has been written is not pertinent to this research, which deals only with item-by-item sequential sampling by attributes. In order to develop the present status of sequential sampling by attributes, it is helpful to discuss attribute sampling generally. To do this requires mention of single, double, and multiple sampling plans. The most widely used sampling plans are probably those of Military Standard 105 and of Dodge-Romig Plans (Gilbreath, 1967).

MILITARY STANDARD 105

Military Standard 105 presents sampling plans that specify an Acceptable Quality Level (AQL), "the maximum percent defective (or the maximum number of defects per hundred units) that, for purposes of sampling inspection, can be considered satisfactory as a process average."[3] These plans represent significance tests of the hypothesis, per cent defective, p', is equal to AQL against the alternate hypothesis, p' is greater than AQL. The maximum number of defects allowed in a sample, C, is established, and the sample size, n, is then determined such that the probability of accepting a lot having p' equal to AQL is equal to $1-\alpha$. Thus, a is the risk of committing an error of type I as in any significance test. No consideration is given to explicitly designating the risk and sampling proceeds on this basis. From each lot of N items, examine a sample of n items. If the sample contains C or fewer defectives, accept the lot; otherwise, reject the lot. The characteristic, α, is not constant in the plans of Military-Standard 105-D but varies with lot size and with the inspection level selected. Matched double and multiple sampling plans are specified in 105-D.

DODGE-ROMIG

In the case of Dodge-Romig plans, the choice of plans is made similarly, except that a constant risk is specified and no consideration is given to α. β, the probability of accepting a lot having $p' = $ Lot Tolerance Per Cent Defective (LTPD) is fixed, the acceptance number C is chosen, and n is determined such that there is a probability of 0.10 of accepting a lot having p' equal to LTPD, and average total inspection (ATI) is minimized. Effective use of Dodge-Romig acceptance sampling procedures requires that rejected lots be screened and all defectives in the sample and in the screened lots be replaced by good items. Other Dodge-Romig plans are provided whereby the Average Outgoing Quality Limit, AOQL, "the maximum of the average outgoing qualities for all possible incoming qualities for a given acceptance sampling plan," is fixed. For both LTPD and AOQL plans, Dodge-Romig plans (Gilbreath, 1967) use minimum ATI as the criterion for determining a specific plan. Summary guide for Dodge-Romig with reference to economy:

> In choosing a value of LTPD (or AOQL) consider and compare the cost of inspection with the economic loss that would ensue if quality as bad as the LTPD were accepted often (or if the average level of quality were greater than the AOQL). Even though the evaluation of economic loss may be difficult, relative values for different levels of per cent defective may often be determined.

Once LTPD is determined, with β fixed at 0.10, "only the least amount of inspection which will accomplish the purpose can be justified." The Dodge-Romig plans are designed to provide minimum ATI and to assure that β is 0.10 when pT is equal to LTPD. Therefore, if β is 0.10 and LTPD was chosen to minimize the costs of accepting bad items and rejecting good ones, Dodge-Romig plans would be least cost plans. No latitude is allowed in selecting β, and no explicit provision is made for minimizing losses by the choice of LTPD (or AOQL). It must be concluded that, in general, Dodge-Romig plans do not provide lease-cost sampling plans. Thus, the plans of Military Standard 105-D assure a high probability of accepting lots of quality AQL or better, and the plans of Dodge-Romig establish a low probability, 0.10, of accepting lots of quality LTPD or poorer. Neither group of plans considers sampling costs and losses explicitly, and neither deals with α and β simultaneously.

Plans Specifying Both α and β

In 1952, J. M. Cameron (Gilbreath, 1967) developed single and double sampling plans using the Poisson approximation to the binomial (the assumed sampling distribution) which established C and n when acceptable quality, p'_1, unacceptable quality, p'_2, α, and β were specified. Once again, the criteria for test design were α and β, and the design was consistent with Neyman and Pearson's hypothesis testing (Gilbreath, 1967). This approach uses a ratio, p_2/p_1, to establish C. When C is known (there are usually two possibilities for C), there may be four plans which approach the desired operating characteristic. For each of the two C values, the designer must adhere strictly to either α or β, whichever is selected, determine n, and allow the other risk to vary slightly. Then the "best" of the four plans as judged by the designer must be chosen for use. Economy may enter into the choice of the best of the four plans, but economy did not enter into the selection of the four plans in the beginning. In 1964, L. A. Johnson (Gilbreath, 1967) surveyed sampling procedures based upon economic and noneconomic criteria, with the following conclusions:

1. Types of inspection schemes presently available provide adequate choice for the form of an acceptance inspection decision rule.
2. Further development of sampling tables based on noneconomic criteria would be of little value.
3. Efforts to utilize economic criteria and knowledge of the process curve to develop inspection procedures have not resulted in any generally applicable theory.
4. There is no general agreement as to the measure of effectiveness for acceptance inspection operations.
5. There is no general agreement as to the appropriate principle of choice in selecting among alternative inspection procedures.
6. Inspection has not been treated adequately as a system of interrelated operations.
7. Successful analysis and design of inspection systems will require a better developed and more clearly understood set of principles – economic, statistical, mathematical, psychological – than now exists relative to acceptance inspection.

After considering the statistical character of acceptance sampling and various measures of the effectiveness of various procedures, it was concluded (Gilbreath, 1967) that monetary units are the superior measures of effectiveness and that Bayes' principle furnishes the best criterion for acceptance of inspection decisions. These conclusions are consistent with A. Hald's work as mentioned previously.

BAYESIAN DECISION RULES

In research at Western Reserve University, Ray E. Schafer (Gilbreath, 1967) agrees that Bayesian statistics provide superior decision rules for categorizing lots in acceptance sampling. However, he argues that the costs relevant to measuring the effectiveness of sampling inspection are so obscure as to be almost unobtainable. Therefore, statistical criteria must be used as measures of effectiveness. Schafer designed single sampling plans by attributes based upon the posterior producer's and consumer's risks for uniform, binomial, hypergeometric, and Polya prior distributions. These are the same prior distributions considered by Hald and Johnson (Gilbreath, 1967). Schafer defines the posterior producer's risk as, "α^*, the posterior conditional probability that, given lot rejection, the number of defective units in the lot was A such that $z \leq z_1$," and the posterior consumer's risk as, "β^*, the posterior conditional probability that, given lot acceptance, the number of defective units, y, in the uninspected portion of the lot is such that $y \geq y_2^*$"

z_1 = A particular value of z such that the producer desires few rejected lots to have $z \leq z_1$.

y_2 = A particular value of y such that the consumer desires few accepted lots to have $y \geq y_2$.

z = x + y, the total number of defectives in a lot of size N.

According to Gilbreath (1967), Hald, Johnson, and Schafer agree that the prior distribution is of utmost importance to the choice of a sampling procedure. Furthermore, they agree that Bayes' Principle is the most intelligent principle of choice.

SEQUENTIAL SAMPLING

In single sampling for attributes, lots of N items are selected for inspection. A sample of n items is drawn from the lot and x, the number of unfavorable attributes in the sample are counted. If x is equal to or less than c, the maximum number of allowable unfavorable attributes for acceptance, the lot is accepted. If x exceeds c, the lot is rejected. The average number of items inspected per lot is a constant. This causes the average cost of sampling inspection per lot to be a constant, regardless of lot quality. Double and multiple sampling provide for reduced average sample sizes. In effect, these plans divide the total sample size into subsamples, each with its own acceptance number. Acceptance or rejection may occur on any sub-sample. Therefore, lots of particularly good or poor quality have a high probability that a decision will be reached on an early sub-sample. Furthermore, if the acceptance number for a subsample is exceeded during inspection of any sub-sample after the first one, rejection may occur without completing inspection of that subsample. This curtailment of inspection further reduces the inspection required, thereby resulting in an additional reduction in average sampling inspection cost. Item-by-item sequential sampling by attributes gives the opportunity to take the maximum advantage of diminished sampling inspection for lots of very good quality or of very poor quality.

ECONOMIC CRITERIA

The use of economic criteria in designing acceptance sampling plans involves the identification of relevant costs and losses, the inclusion of these economic parameters in a sampling model, and the formulation of a decision rule that attempts to minimize the total expected cost of sampling inspection. Costs associated with inspection and decisions should be the design criteria for sampling inspection plans. There is a difficulty of obtaining accurate costs for such conditions as acceptance of bad lots and rejection of good ones, but nevertheless these are the losses, along with inspection costs, producers and consumers are trying to reduce by sampling. Three costs affiliated with sampling inspection are identified:

1. $$\frac{\text{Average cost of accepting a defective item}}{\text{Average cost of accepting a defective item}} = 1$$

2. Case I, sorting, $Kr = \dfrac{Sorting\ cost/item}{Average\ cost\ of\ accepting\ a\ defective\ item}$

Case II, non-sorting, $Kr = \dfrac{Manufactureing\ cost/item}{Average\ cost\ of\ accepting\ a\ defective\ item}$

3. $ks = \dfrac{Sampling\ and\ testing\ cost\ per\ item\left(In\ Sample\right)}{Average\ cost\ of\ accepting\ a\ defective\ item}$

APPLICATION OF DECISION THEORY

The literature contains many ideas about the design of sampling procedures using statistical decision theory. The sampling inspection of mass-produced items is a field of special application for the theory of statistical decisions, which has received such an impetus from the work of many researchers. In the theory of statistical decisions, we envisage a situation where the following occur:

(i)   A choice must be made from among a series of possible actions, on the basis of data subject to random variations.
(ii) It is possible to associate a definite measure of loss to each possible action when it is inappropriately taken, and
(iii) The choice in question can legitimately be regarded as one of a series of statistically independent choices, so that it is reasonable to act in each case in such a way as to minimize the expected value of the loss.

Situations of this sort rarely occur in practical life, where one or more of the three essential features are usually missing. It may not be possible to enumerate clearly in advance all the alternative courses of action; or it may not be possible to associate each alternative with a definite measure of loss. Alternatively, the choice in question may have a unique character which makes it inappropriate to regard it as one of a series. But in the case of sampling inspection of batches of items we do little damage to the true nature of the practical problem if we regard it as coming within the scope of the theory. The alternative courses of action, acceptance of the batch, rejection, acceptance subject to further inspection, down-grading, etc., are fairly clear cut; quite good estimates can often be made of the monetary losses arising from wrong decisions; and routine cases, by definition, can be regarded as forming a series, and each decision can be regarded as final.

There are many problems associated with the use of cost criteria and Bayesian decision rules in sampling inspection. Among those are:

(1) The importance of the form of the prior distribution.
(2) Mood's theorem concerning the pure binomial prior distribution.
(3) Evidence for the mixed binomial prior distribution.
(4) The dependence of sample size on lot or batch size.
(5) Applicability of other principles of choice.

Knowledge of the true form of the process curve or prior distribution is very important when applying statistical decision theory in sampling inspection. Gilbreath (1967) echoed G. E. P. Box's comment that it is usually a relatively simple matter to calculate the loss in wrongly rejecting a good batch. The difficulty arises in deciding the cost of wrongly accepting a bad batch since this is tied up with such difficult questions as the value to a manufacturer of a customer's goodwill. As we can note nowadays, in the wake of the COVID-19 pandemic, that the manufacturer's goodwill with the customer comes into plan in reacting to supply chain disruptions.

OLIVER-BADIRU EXPERT SYSTEM FOR SUPPLIER DEVELOPMENT

Over three decades ago, Oliver and Badiru (1993) developed an expert system as a decision support system for supply chain management. Specifically, they developed an expert system for supplier development programs (SDPs) in manufacturing companies. The productive capacity of a manufacturer depends on the viability of its interfaces with its suppliers. A SDP refers to a continuous process of identifying, assessing, classifying, supporting, and developing suppliers with an objective to create and maintain a network of competent and consistent suppliers.

Organizational excellence would be better off if we could retain all the experts in various fields. Recognizing the limitations imposed by nature, Oliver and Badiru (1993) conducted research on knowledge acquisition, encoding, and dissemination. This implies the efficacy of expert systems. The Oliver-Badiru SDP software captures human expertise in the domain of supply chain. No manufacturing firm is totally self-sufficient. All firms depend on external suppliers to meet a variety of needs in the form of materials, goods, or services. On an average, manufacturers in North America spend over 60% of total revenue on acquisition of goods and supplies from outside suppliers. Factors like productivity, competitiveness, quality, and cost of products of a firm are affected by the performance of suppliers. With increasing global competition and the need for continuous improvement, the need for good suppliers is critical, particularly following an unfortunate disruption, such as what was caused by the COVID-19 pandemic in the year 2020. Having good suppliers is similar to having good business partners.

Oliver-Badiru SDP involves a variety of parameters, including quality, price, delivery, technology, geographical location, political environment, organizational policies, situational requirements, and other factors, attributes, and indicators, which further branch into numerous other considerations with differing levels of detail. Like other areas of materials management, it is a scenario of incompleteness, inconsistency, inaccuracy, uncertainty of data or relations, unstructured, complex, and time-dependent procedures. The activities involved are hard or impossible to describe by quantitative tools, such as mathematical algorithms and repetitive processes. Managers in SDPs use a combination of facts, heuristic, and experience to tackle problems. All these are encoded in the Oliver-Badiru SDP software (Oliver and Badiru, 1993). In the 1990s, there was a lack of established methodologies for SDPs. The situation has vastly improved in recent years due to the emergence of more sophisticated computer hardware and software. Expert Systems, which form a branch of artificial intelligence, have been the most productive and promising applications of AI in business and industry today (Badiru, 2021). The methodology of the SDP software has the following capabilities:

• Performs organizational assessment to detect if there is a need for SDP.
• Evaluates and rates suppliers based on the parameters defined and weights determined.
• Classifies suppliers into five groups based on ratings and user consultation.
• Offers advice on how to improve a supplier's performance.

Organizational Assessment for SDP refers to analyzing the existing situation in the company and the suppliers, to detect if there is a need for SDP. A firm might need SDP due to one or more of the drivers listed below:

• Establish a source for the new product or part
• Available supplier unwilling to supply
• Improve quality of incoming material
• Improve delivery performance
• Improve service
• Reduce cost of material
• Improve technical capability
• Reduce supplier base
• Enlarge supplier base
• Meet social, political, geographical, and environmental concerns
• Overcome market deficiency
• Change in the firm's operating systems (e.g., Introduction of JIT)
• Large future requirement

The Organizational Assessment module of the model has embedded rules which focus on meeting the objectives mentioned above. From user responses to questions about the above-mentioned objectives and comparison with the existing situation, the expert system makes recommendations. An example of a rule in SDP is presented below:

```
IF {Firm plans to introduce JIT: Just In Time}
THEN {SDP is recommended - Focus on JIT parameters
during evaluation}
```

Using if-then-else rules within the expert system, the list of parameters embodied in SDP is summarized below:

QUALITY IMPROVEMENT

1. Percentage of orders meeting design specification
2. Percentage of in-process – defects detected
3. Percentage of downtime resulting from supplier error
4. Quality of raw materials or parts used by supplier
5. Supplier following world class quality standards
6. Participating In the firm's qualify program

COST OF PRODUCT

1. Percentage of difference between actual and budgeted price
2. Price history (Increase or decrease)
3. Number of cost reduction programs initiated by supplier
4. Worldwide competitive cost structure
5. Credit rating

DELIVERY COMMITMENTS

1. Percentage of on-time deliveries
2. Percentage of orders meeting quantity requirements
3. Average time for normal orders
4. Average time for special and rush orders
5. Incremental delivery of a large order
6. Frequent deliveries
7. Ability to accommodate current volume

CUSTOMER SERVICE

1. Product and service availability
2. Personnel capabilities
3. Time required to correct supplier error
4. Prompt and accurate reply to communications
5. Advance written notice of any price change
6. Advance notification about any delivery change

TECHNICAL CAPABILITIES

1. Manufacturing process
2. Advanced facilities like CAD/CAM
3. Cycle time reduction for new products and process
4. Number of product improvements initiated by suppliers
5. Engineering and product development support

OPERATIONAL REQUIREMENTS

1. A common coding system
2. Packaging, labeling, and shipping requirements
3. Proper paper work
4. Supplier ability to handle paper-based transactions
5. Billing errors

STRATEGIC ALLIANCE

1. Long-term co trunk meet for productivity and quality improvement
2. Growth of the company
3. Supplier's financial ability
4. Ability to accommodate forecasted volume requirements
5. Supplier's willingness to dedicate capacity, resources
6. Coordinate flow of data meeting objective of both organizations
7. Cooperate with firm's supplier evaluation and rating system
8. Respect confidentiality of boldness
9. Respect the terms and conditions of the contract
10. Has Electronic Data Interchange system
11. Supplier ability to service JIT
12. Early supplier development
13. Suggest better or less expensive ways of using supplier product
14. Considers the firm as a top customer
15. Geographical proximity
16. Social reasons like environment

There are several parameters considered by managers in a SDP program to evaluate suppliers. These differ between organizations, products, and suppliers. Analysis and classification of these parameters involved is vital for developing a general model of SDP. The parameters involved are identified and classified during the development of SDP based on the prevailing methods available in the literature references of the 1990s. In the Categorical Method, for each parameter, the user assigns a (+), (-), or (0) for preferred, unsatisfactory, and neutral, respectively. This method is not effective, since all the parameters are weighted equally. Weighted-Point method assigns weights to the parameters, but all the performance measures act in uniform units, which is percentage. In reality, it is not possible to quantify many parameters to percentage. The Cost-Ratio method calculates cost ratios for several parameters, for different suppliers. It requires a comprehensive cost-accounting system. Dimensional Analysis does numerical comparison of vendor performance with company standards and calculates an index based on weights assigned. But it is hard to convert many subjective data to numeric. The Analytic Hierarchy Process (AHP) has also been used to model supplier evaluation (Oliver and Badiru, 1993). The AHP requires pairwise comparison of all the suppliers for all the parameters. This is a tedious process when there is a large number of parameters involved. The evaluation technique used in SDP involves three steps:

(1) Assigning weights to the parameters using the AHP
(2) User selected rating on a five-point scale (0–4) that best characterizes a supplier with aspect to each parameter
(3) Computing the final rating for each supplier

Assigning Weights

Initialization of weights to the parameters is done using the AHP. The AHP involves Pairwise comparison of parameters with respect to their importance. Pairwise comparison is done separately for the classification titles and for the set of parameters under all classifications. The procedure involved in the AHP is explained with the following example, computing the relative weights of the classification titles:

(1) Strategic Alliance
(2) Quality Improvement
(3) Delivery Commitments
(4) Customer Service
(5) Technical Capability
(6) Cost of Products
(7) Operational Requirements

Pairwise comparisons of parameters are done for "importance to the firm." A matrix of pairwise comparisons (not shown here) is made based on user response to questions like the example presented below.

> **Question**: With respect to importance to your firm how do you compare Strategic Alliance with Quality Improvement?
> **Options**: Absolutely Less Important (Weight: 1/9), Slightly Less Important (Weight: 1/5), Equally Important (Weight: 1), Slightly More Important (Weight: 5), and Absolutely More Important (Weight: 9).

Once the pairwise comparison is complete, normalization is done by dividing each entry in a column by the sum of all the entries in the column and the normalized average weights associated with each classification title are calculated. The AHP is similarly done for all parameters, under all classifications, comparing with the other parameters in the classification. Thus, values for the following variables are obtained:

1. C_i – Weight of title of classification
2. P_{ij} – Weight of parameter

where i = 1 to L (Number of classifications)
j = 1 to M (Number of parameters under a classification)

Rating Suppliers

The rating module of the ES model presents the user with questions about all parameters. The user selects the best option among five which characterizes the supplier under evaluation. For example,

Question: With respect to quality of raw materials or parts used by supplier, select the option which best characterizes the supplier.
Choices: Poor(0), Fair(1), Good (2), Very Good (3), Excellent (4).

The terminology used for choices in different questions for each parameter will differ, but the points assigned will remain the same, from 0 to 4 in all cases. This gives the provision to assign zero points to suppliers who do not qualify at all to score any points for a particular parameter. The rating process is repeated for all the parameters and for all the suppliers. Thus, values for the variables, S_{ijk}, where $k = 1$ to N, (Number of suppliers) are found. The final rating R_k of a supplier is computed using the formula shown below:

$$R_k = \sum_{i=1}^{L} \sum_{j=1}^{M} C_i P_{ij} S_{ijk}$$

Even though there are different numbers of parameters under each classification, since they are multiplied by the weights of the parameters, they are normalized and vary from 0 to 4. The maximum rating possible for a supplier is 4 and the minimum rating possible is 0. For an illustrative example of the SPD application to a practical supplier case example, please see Oliver and Badiru (1993).

SUPPLIER CLASSIFICATION

Based on the ratings obtained for each supplier, they are classified into five groups – (1) Partner, (2) Alliance, (3) Annual Renewal Supplier, (4) Normal Supplier, and (5) Unacceptable Supplier. The required rating for each classification is decided by user.

SUPPLIER ADVISING

Supplier advising is the process of informing the suppliers about their classification, reasons for the classifications, and suggestions on how they can improve. This module performs based on embedded rules which consist of advice for lack of specific parameters. A sample rule is as follows:

```
IF {Supplier has no ED I Facility}
    AND {Firm's objective is to implement EDI}
    THEN {Recommend: Get additional from plant manager
```

In a scenario of traditional buyer–supplier relationship being replaced with partner–partner relationship, SDP is vital for organizations to remain competitive. SDP involves many subjective but strategically pertinent data. Hence, an expert system model is useful for capturing and representing real-world scenarios. The parameters listed for organizational assessment and supplier evaluation are useful for organizations to identify specific parameters for their SDPs. This general Oliver-Badiru SDP model is adaptable for many situations and is suitable for modification to meet specific needs, particularly in the modern era of digital and global supply chain networks. Having a strong SDP can help retain, enhance, and prolong supplier relationships.

OTHER LITERATURE EXAMPLES OF SUPPLY CHAIN APPLICATIONS

As mentioned earlier in this chapter and the preceding chapter, IE has a lot to offer in terms of tools and techniques applied to supply chain research, practice, and applications. The literature abounds with several examples. Loska and Higa (2020) address supply chain risk management for the future of organic supply chain for the US Air Force. Ekstrom et al. (2020) present differentiation strategies for defense supply chain design, with a specific focus on the Swedish defense system. Ellis et al. (2018) address the rising operational costs and software sustainment concerns in the US Air Force with respect to moving to newer technology for the Air Force standard base supply chain. With a base in the Netherlands, Van Strien et al. (2019) present a risk-management approach to performance-based contracting in military supply chains. The paper investigates factors that influence service provider's willingness to accept risks induced by performance-based contracting. Food supply and food security, topics of particular focus for industrial engineers, are addressed by Lin et al. (2019) and Christensen et al. (2021). Overstreet et al. (2019) introduce a multi-study analysis of learning culture, human capital, and operational performance in supply chain management. The study empirically evaluates the relationship between learning culture, workforce level, human capital, and operational performance in two diverse supply chain populations, aircraft maintenance, and logistics readiness. This study is particularly of interest in this book because of its learning curve alignment with the contents of Chapter 5. The contents addressing human capital and workforce level are also of special interest in relation to the earlier-era practice of People Off Payroll (POP) by ill-informed manufacturers. People are critical to the success of any supply chain. The COVID-19 pandemic lockdowns around the world make this fact painfully obvious. After the pandemic, employers found it difficult to rehire, retrain, and retain employees that were retrenched due to the pandemic. A lot of learning curve assets were, consequently, lost. Even for those who came back to work, the learn-forget-relearn processes (Badiru, 2015) came into play.

Within the realm of advanced research studies, Morrow (2021) and Femano (2021) present doctoral dissertation reports on the supply chain. According to Morrow (2021), information is crucial to supply chain performance because it is used to make decisions and trigger actions. Organizations across world-class supply chains increasingly use information technology to analyze and process supply chain data. However, supply chain management lacks a common language, making information exchange difficult. An ontology can provide a standardized framework that organizes a given knowledge domain. Morrow's research proposes a common language or ontology for supply chain management that can be understood by both humans and computers. This is an example of a human-systems integration research for supply chains. According to current research, an established and widely used supply chain framework is a good starting point for developing a supply chain ontology. Many researchers recommend using the Supply Chain Operations Reference (SCOR) Model. This framework is translated into a software package that generates a Web Ontology Language (OWL), which can be used by information technology. Using the SCOR 12.0 as the framework, an XML/OWL-based model, Morrow developed a tool, which can be used by information technology to improve information exchanges between supply chain partners.

In his own doctoral research, Femano (2021) emphasizes that businesses operate every day in a disruptive environment. Supply and demand uncertainty, natural disasters, global pandemics, and mishaps can all cause chaos to a supply chain's flow. It is impossible to predict every disruption a supply chain may encounter. The best an organization can do to protect network performance is to build resilience in the supply chain and lifeline of its operations. Ensuring that a supply chain has the proper built-in mechanisms to resist and recover from disruptions is referred to as Supply Chain Resilience (SCR) (Femano, 2021). While it is generally agreed that SCR can be improved through the implementation of SCR strategies, the links between these strategies, performance improvement, and resilience is understudied. Femano's research focuses on resource-based view and theory of constraints to categorize the SCR strategies, examine the links between the strategies and performance, and develop a metric to measure network resilience over time. First, a meta-analytical study identifies generalizable relationships between SCR strategies and the organization's performance measures. Then, the SCR redundancy strategies are applied to a model simulation to illustrate the resilience curve response to different SCR strategic decisions. Resilience outcomes are compared using a developed Resilience Capability Metric (RCM) utilizing Area under the Curve to measure the cumulative performance level of the system from disruption to a predetermined endpoint, representing how much of the system demand can be served by different network resilience designs. Finally, SCR flexibility strategies are analyzed to see how constraints imposed on a supply chain's response time could impact the resilience of the supply chain. Femano's work highlights the positive impact on performance and resilience that can be realized when organizations take the time to implement the proper supply chain resilience strategies, while providing managers with RCM to measure and compare the impact of different strategies within their organization.

Other relevant topics and ideas, pertinent for IE applications to the supply chain, are presented by Sinha et al. (2020), Perez (2014), Dolgui et al. (2018), Lin et al. (2019), Reynolds (2017), and Hernandez-Espallardo et al. (2010). A good illustration of the diverse application of IE to the supply chain of diverse industries is presented by Barve and Shanker (2020), in which they address Green Supply Chain Management. Their study focuses on detecting various parameters associated with green supply chain management practices in diamond mining industries globally.

CONCLUSION

As demonstrated by the narratives, examples, and research reports presented in this chapter, supply chain research, management, and application are within the operational spectrum of IE. Of particular relevance are the human-centric aspects of the practice of IE. The chapters that follow present a mix of quantitative and qualitative tools and techniques for supply chain analysis, management, and control.

REFERENCES

Badiru, A. B. (2014), editor; *Handbook of Industrial & Systems Engineering*, 2nd Edition, Taylor & Francis Group/CRC Press, Boca Raton, FL.

Badiru, A. B. (2015), "Quality Insights: Learning, forgetting, and relearning quality: a half-life learning curve modeling approach," *International Journal of Quality Engineering and Technology*, Vol. 5, No. 1, pp. 79–100.

Badiru, A. B. (2019), *Systems Engineering Models: Theory, Methods, and Applications*, Taylor & Francis Group/CRC Press, Boca Raton, FL.

Badiru, A. B. (2021), *Artificial Intelligence and Digital Systems Engineering*, Taylor & Francis Group/CRC Press, Boca Raton, FL.

Barve, Akhelesh and Saket Shanker (2020), "Sustainable Supply Chain Concerns in Diamond Industries," *Proceedings of the International Conference on Industrial Engineering and Operations Management*, Dubai, UAE, March 10–12, 2020.

Christensen, Cade, Torrey Wagner, and Brent Langhals (2021), "Year-Independent Prediction of Food Insecurity Using Classical and Neural Network Machine Learning Methods," *AI*, Vol. 2, No. 1, pp. 244–260. doi:10.3390/ai2020015

Dolgui, A., Ivanov, D., and Sokolov, B. (2018), "Ripple effect in the supply chain: an analysis and recent literature," *International Journal of Production Research*, 56(1-2), 414–430.

Ekstrom, Thomas, Per Hilletofth, and Per Skoglund (2020), "Differentiation strategies for defense supply chain design," *Journal of Defense Analytics and Logistics*, Vol. 4, No. 2, pp. 183–202: https://www.emerald.com/insight/2399-6439.htm

Ellis, Tommie L., Robert A. Nicholson, Antoinette Y. Briggs, Scott A. Hunter, James E. Harbison, Paul S. Saladna, Michael W. Garris, Robert K. Ohnemus, John E. O'Connor, and Steven B. Reynolds (2018), "Lifting and shifting the Air Force retail supply system," *Journal of Defense Analytics and Logistics*, Vol. 1, No. 2, pp. 172–184.

Femano, Amanda L. (2021), "Performance Improvement Through Better Understanding of Supply Chain Resilience," Ph.D. Dissertation, Air Force Institute of Technology, Wright-Patterson Air Force Base, Ohio, 2021.

Gilbreath, Sidney Gordon (1967), A Bayesian Procedure for the Design of Sequential Sampling Plans, Ph.D. Dissertation, School of Industrial Engineering, Georgia Institute of Technology, Atlanta, GA, 1967.

Hernández-Espallardo, Miguel, Augusto Rodríguez-Orejuela, and Manuel Sánchez-Pérez (2010), "Inter-organizational governance, learning and performance in supply chains," *Supply Chain Management: An International Journal*, Vol. 15, No. 2, pp. 101–114 Permanent link to this document: doi: 10.1108/13598541011028714

Lin, Xiaowen, Paul J Ruess, Landon Marston, and Megan Konar (2019), "Food flows between counties in the United States," *Environmental Research Letters*, Vol. 14, No. 2, pp. 2–18, doi: 10.1088/1748-9326/ab29ae

Loska, David and James Higa (2020), "The risk to reconstitution: supply chain risk management for the future of the US Air Force's Organic Supply Chain," *Journal of Defense Analytics and Logistics*, Vol. 4, No. 1, pp. 19–40: https://www.emerald.com/insight/2399-6439.htm

Morrow, David (2021), "Developing A Basic Formal Supply Chain Ontology to Improve Communication and Interoperability," Ph.D. Dissertation, Air Force Institute of Technology, Wright-Patterson Air Force Base, Ohio, 2021.

Oliver, Gaugarin E. and A. B. Badiru, "An Expert System Model for Supplier Development Program in a Manufacturing Firm," *Proceedings of the 7th Oklahoma Symposium on Artificial Intelligence, Stillwater*, Oklahoma, November 18–19, 1993, pp. 135–141.

Overstreet, Robert E., Joseph B. Skipper, Joseph R. Huscroft, Matt J. Cherry, and Andrew L. Cooper (2019), "Mult-study analysis of learning culture, human capital, and operational performance in supply chain management: The moderating role of workforce level," *Journal of Defense Analytics and Logistics*, Vol. 3, No. 1, pp. 41–59: https://www.emerald.com/insight/2399-6439.htm

Perez, Hernan David (2014), *Supply Chain Roadmap: Aligning Supply Chain with Business Strategy*, www.SupplyChainRoadmap.com, printed by David Hernan Perez, Charleston, NC.

Reynolds (2017), "Lifting and shifting the Air Force retail supply system," *Journal of Defense Analytics and Logistics*, Vol. 1, No. 2, 2020, pp. 19–40: https://www.emerald.com/insight/2399-6439.htm

Sinha, Amit, Ednilson Bernardes, Rafael Calderon, and Thorsten Wuest (2020), *Digital Supply Networks*, McGraw-Hill, New Corlk, NY.

Van Strien, Jeroen, Cees Johannes Gelderman, and Janjaap Semeijn (2019), "Performance-based contracting in military supply chains and the willingness to bear risks," *Journal of Defense Analytics and Logistics*, Vol. 3, No. 1, pp. 83–107: https://www.emerald.com/insight/2399-6439.htm

3 Forecasting and Inventory Modeling for Supply Chains

FORECASTING FOR THE SUPPLY CHAIN

Any supply chain is subject to dynamic changes in the production, shipment, and delivery processes. For the supply to be adaptive, responsive, and resilient to the changes, either expected or unexpected, a combination of analytical, qualitative, and computer techniques must be employed. Good forecasting is the basis for achieving a responsive supply chain. Hogan et al. (2020), Badiru et al. (1993), Badiru (2019), and references therein present tools and techniques pertinent for forecasting in the supply chain environment.

Managing complex supply chains effectively calls for good information, which can be provided by forecasting and inventory control. Forecasting is not just for marketing and production planning purposes. This chapter presents techniques of forecasting and inventory management as a part of the overall quantitative techniques for supply chain planning, design, and control. Several analytical tools are essential for analyzing project systems. These are relevant if we formulate a supply chain as a project system as discussed in Chapter 1. Prior to proceeding to the project management phase, a good understanding of the enterprise system, within which the supply chain resides, is indispensable for getting a successful output.

Forecasting should be an important part of overall supply strategy. Effective prediction provides information needed to make good enterprise-wide decisions. Several techniques are available for forecasting. Regression, time series analysis, computer simulation, and artificial neural networks are common examples of forecasting techniques. There are two basic types of forecasting: *intrinsic forecasting* and *extrinsic forecasting*.

Intrinsic forecasting is based on the assumption that historical data adequately describe the problem scenario to be forecasted. Forecasting models based on historical data require extrapolation to generate estimates for the future. The requirements of intrinsic forecasting are:

- Collect historical data.
- Develop quantitative forecasting model based on the data collected.
- Generate forecasts recursively for the future.
- Revise the forecasts as new pieces of data become available.

DOI: 10.1201/9781003111979-3

Extrinsic forecasting assumes that the forecasts to be generated are correlated to some other external factors such that the forecasts of the external factors provide reliable forecasts for the current problem. For example, the demand for a new product may be based on forecasts of household incomes. Before any forecasting system is implemented, a complete analysis of the data required must be performed. This is useful for setting activity times and task allocation strategies.

DATA MEASUREMENT SCALES FOR FORECASTING

Forecasting requires data collection, measurement, and analysis. In the supply chain, the analyst will encounter different types of measurement scales depending on the particular items involved. Data may need to be collected on shipment schedules, costs, performance levels, problems, and so on. The different types of data measurement scales that are applicable are presented below.

Nominal scale of measurement: A *nominal scale* is the lowest level of measurement scales. It classifies items into categories. The categories are mutually exclusive and collectively exhaustive. That is, the categories do not overlap and they cover all possible categories of the characteristics being observed. For example, in the analysis of the critical path in a project network, each job is classified as either critical or not critical. Gender, type of industry, job classification, and color are some examples of measurements on a nominal scale.

Ordinal scale of measurement: An *ordinal scale* is distinguished from a nominal scale by the property of order among the categories. An example is the process of prioritizing project tasks for resource allocation. We know that first is above second, but we do not know how far above. Similarly, we know that better is preferred to good, but we do not know by how much. In quality control, the ABC classification of items based on the Pareto distribution is an example of a measurement on an ordinal scale.

Interval scale of measurement: An *interval scale* is distinguished from an ordinal scale by having equal intervals between the units of measure. The assignment of priority ratings to project objectives on a scale of 0–10 is an example of a measurement on an interval scale. Even though an objective may have a priority rating of 0, it does not mean that the objective has absolutely no significance to the project team. Similarly, the scoring of 0 on an examination does not imply that a student knows absolutely nothing about the materials covered by the examination. Temperature is a good example of an item that is measured on an interval scale. Even though there is a zero point on the temperature scale, it is an arbitrary relative measure. Other examples of interval scales are IQ measurements and aptitude ratings.

Ratio scale of measurement: A *ratio scale* has the same properties of an interval scale but with a true zero point. For example, an estimate of a zero time unit for the duration of a task is a ratio scale measurement. Other examples of items measured on a ratio scale are cost, time, volume, length, height, weight, and inventory level. Many of the items measured in a project management environment will be on a ratio scale.

Another important aspect of data analysis for project control involves the classification scheme used. Most projects will have both *quantitative* and *qualitative* data. Quantitative data require that we describe the characteristics of the items being studied numerically. Qualitative data, on the other hand, are associated with object attributes that are not measured numerically. Most items measured on the nominal and ordinal scales will normally be classified into the qualitative data category, while those measured on the interval and ratio scales will normally be classified into the quantitative data category.

The implication for project control is that qualitative data can lead to bias in the control mechanism because qualitative data are subject to the personal views and interpretations of the person using the data. Whenever possible, data for project control should be based on quantitative measurements.

There is a class of project data referred to as *transient data*. This is defined as a volatile set of data that is used for one-time decision-making and is not then needed again. An example may be the number of operators that show up at a job site on a given day. Unless there is some correlation between the day-to-day attendance records of operators, this piece of information will have relevance only for that given day. The project manager can make his decision for that day on the basis of that day's attendance record. Transient data need not be stored in a permanent database unless it may be needed for future analysis or uses (e.g., forecasting, incentive programs, and performance review).

Recurring data refer to data that are encountered frequently enough to necessitate storage on a permanent basis. An example is a file containing contract due dates. This file will need to be kept at least through the project life cycle. Recurring data may be further categorized into *static data* and *dynamic data*. Recurring data that are static will retain their original parameters and values each time they are retrieved and used. Recurring data that are dynamic have the potential for taking on different parameters and values each time they are retrieved and used. Storage and retrieval considerations for project control should address the following questions:

1. What is the origin of the data?
2. How long will the data be maintained?
3. Who needs access to the data?
4. What will the data be used for?
5. How often will the data be needed?
6. Are the data for look-up purposes only (i.e., no printouts)?
7. Are the data for reporting purposes (i.e., generate reports)?
8. In what format are the data needed?
9. How fast will the data need to be retrieved?
10. What security measures are needed for the data?

DATA DETERMINATION AND COLLECTION

It is essential to determine what data to collect for project control purposes. Data collection and analysis are basic components of generating information for project control. The requirements for data collection are discussed next.

Choosing the data: This involves selecting data on the basis of their relevance and the level of likelihood that they will be needed for future decisions and whether or not they contribute to making the decision better. The intended users of the data should also be identified.

Collecting the data: This identifies a suitable method of collecting the data as well as the source from which the data will be collected. The collection method will depend on the particular operation being addressed. The common methods include manual tabulation, direct keyboard entry, optical character reader, magnetic coding, electronic scanner, and more recently, voice command. An input control may be used to confirm the accuracy of collected data. Examples of items to control when collecting data are the following:

Relevance check: This checks if the data are relevant to the prevailing problem. For example, data collected on personnel productivity may not be relevant for a decision involving marketing strategies.

Limit check: This checks to ensure that the data are within known or acceptable limits. For example, an employee's overtime claim amounting to over 80 hours/week for several weeks in a row is an indication of a record well beyond ordinary limits.

Critical value: This identifies a boundary point for data values. Values below or above a critical value fall in different data categories. For example, the lower specification limit for a given characteristic of a product is a critical value that determines whether or not the product meets quality requirements.

Coding the data: This refers to the technique used in representing data in a form useful for generating information. This should be done in a compact and yet meaningful format. The performance of information systems can be greatly improved if effective data formats and coding are designed into the system right from the beginning.

Processing the data: Data processing is the manipulation of data to generate useful information. Different types of information may be generated from a given data set depending on how it is processed. The processing method should consider how the information will be used, who will be using it, and what caliber of system response time is desired. If possible, processing controls should be used. This may involve:

Control total: Check for the completeness of the processing by comparing accumulated results to a known total. An example of this is the comparison of machine throughput to a standard production level or the comparison of cumulative project budget depletion to a cost accounting standard.

Consistency check: Check if the processing is producing the same results for similar data. For example, an electronic inspection device that suddenly shows a measurement that is ten times higher than the norm warrants an investigation of both the input and the processing mechanisms.

Scales of measurement: For numeric scales, specify units of measurement, increments, the zero point on the measurement scale, and the range of values.

Using the information: Using information involves people. Computers can collect data, manipulate data, and generate information, but the ultimate decision rests with people, and decision-making starts when information becomes available. Intuition, experience, training, interest, and ethics are just a few of the factors that determine how people use information. The same piece of information that is positively used to further the progress of a project in one instance may also be used negatively in another instance. To assure that data and information are used appropriately, computer-based security measures can be built into the information system.

Project data may be obtained from several sources. Some potential sources are:

- Formal reports
- Interviews and surveys
- Regular project meetings
- Personnel time cards or work schedules

The timing of data is also very important for project control purposes. The contents, level of detail, and frequency of data can affect the control process. An important aspect of project management is the determination of the data required to generate the information needed for project control. The function of keeping track of the vast quantity of rapidly changing and interrelated data about project attributes can be very complicated. The major steps involved in data analysis for project control are:

- Data collection
- Data analysis and presentation
- Decision making
- Implementation of action

Data are processed to generate information. Information is analyzed by the decision maker to make the required decisions. Good decisions are based on timely and relevant information, which in turn is based on reliable data. Data analysis for project control may involve the following functions:

- Organizing and printing computer-generated information in a form usable by managers.
- Integrating different hardware and software systems to communicate in the same project environment.

- Incorporating new technologies such as expert systems into data analysis.
- Using graphics and other presentation techniques to convey project information.

Proper data management will prevent misuse, misinterpretation, or mishandling. Data are needed at every stage in the life cycle of a project from the problem identification stage through the project phase-out stage. The various items for which data may be needed are project specifications, feasibility study, resource availability, staff size, schedule, project status, performance data, and phase-out plan. The documentation of data requirements should cover the following:

Data summary: A data summary is a general summary of the information and decision for which the data are required as well as the form in which the data should be prepared. The summary indicates the impact of the data requirements on the organizational goals.

Data processing environment: The processing environment identifies the project for which the data are required, the user personnel, and the computer system to be used in processing the data. It refers to the project request or authorization and relationship to other projects and specifies the expected data communication needs and mode of transmission.

Data policies and procedures: Data handling policies and procedures describe policies governing data handling, storage, and modification and the specific procedures for implementing changes to the data. Additionally, they provide instructions for data collection and organization.

Static data: A static data description describes that portion of the data that are used mainly for reference purposes and it is rarely updated.

Dynamic data: A dynamic data description describes that portion of the data that are frequently updated based on the prevailing circumstances in the organization.

Data frequency: The frequency of data update specifies the expected frequency of data change for the dynamic portion of the data, for example, quarterly. This data change frequency should be described in relation to the frequency of processing.

Data constraints: Data constraints refer to the limitations on the data requirements. Constraints may be procedural (e.g., based on corporate policy), technical (e.g., based on computer limitations), or imposed (e.g., based on project goals).

Data compatibility: Data compatibility analysis involves ensuring that data collected for project control needs will be compatible with future needs.

Data contingency: A data contingency plan concerns data security measures in case of accidental or deliberate damage or sabotage affecting hardware, software, or personnel.

FORECASTING BASED ON AVERAGES

The most common forecasting techniques are based on averages. Sophisticated quantitative forecasting models can be formulated from basic average formulas. The traditional methods of averages are presented as follows:

Simple Average Forecast

In this method, the forecast for the next period is computed as the arithmetic average of the preceding data points. This is often referred to as average to date. That is,

$$f_{n+1} = \frac{\sum_{t=1}^{n} d}{n}$$

where

f_{n+1} is the forecast for period $n + 1$
d is the data element for the period in question
n is the number of preceding periods for which data are available

Period Moving Average Forecast

In this method, the forecast for the next period is based only on the most recent data values. Each time a new value is included, the oldest value is dropped. Thus, the average is always computed from a fixed number of values. This is represented as

$$f_{n+1} = \frac{\sum_{t=n-T+1}^{n} d_t}{T}$$
$$= \frac{d_{n-T+1} + d_{n-T+2} + \cdots + d_{n-1} + d_n}{T}$$

where

f_{n+1} is the forecast for period $n + 1$
d_t is the datum for period t
T is the number of preceding periods included in the moving average calculation
n is the current period at which forecast of f_{n+1} is calculated

The moving average technique is an after-the-fact approach. Since T data points are needed to generate a forecast, we cannot generate forecasts for the first $T - 1$ periods. But this shortcoming is quickly overcome as the number of data points available becomes large.

Weighted Average Forecast

The weighted average forecast method is based on the assumption that some data points might be more significant that others in generating future forecasts. For example, the most recent data points may weigh more than very old data points in the calculation of future estimates. This is expressed as

$$f_{n+1} = \frac{\sum_{t=1}^{n} w_i d_t}{\sum_{t=1}^{n} w_t}$$

$$= \frac{w_1 d_1 + w_2 d_2 + \cdots + w_n d_n}{w_1 + w_2 + \cdots + w_n}$$

where

f_{n+1} is the weighted average forecast for period $n + 1$

d_t is the datum for period t

T is the additional notation representing the planning horizon for the forecast problem

n is the current period at which forecast of f_{n+1} is calculated

w_t is the weight of data point t

The w_ts are the respective weights of the data points such that

$$\sum_{t=1}^{n} w_t = 1.0$$

Weighted T-Period Moving Average Forecast

In this technique, the forecast for the next period is computed as the weighted average of past data points over the last T time periods. That is,

$$f_{n+1} = w_1 d_n + w_2 d_{n-1} + \cdots + w_T d_{n-T+1}$$

where w_is are the respective weights of the data points such that

$$\sum_{i=1}^{n} w_i = 1.0$$

Exponential Smoothing Forecast

This is a special case of weighted moving average forecast. The forecast for the next period is computed as the weighted average of the immediate past data point and the forecast of the previous period. In order words, the previous forecast is adjusted based on the deviation (forecast error) of that forecast from the actual data. That is,

$$f_{n+1} = \alpha d_n + (1-\alpha) f_n$$
$$= f_n + \alpha (d_n - f_n)$$

where
f_{n+1} is the exponentially weighted average forecast for period $n + 1$
d_n is the datum for period n
f_n is the forecast for period n
α is the smoothing factor (real number between 0 and 1)

A low smoothing factor gives a high degree of smoothing, while a high value causes the forecast to closely match actual data.

REGRESSION ANALYSIS

The primary function of regression analysis is to develop a model which expresses the relationship between a dependent variable and one or more independent variables. It is sometimes called line fitting or curve fitting. Regression analysis is an important statistical tool that can be applied to many prediction and forecasting problems in the project environment. The utility of a regression model is often tested by analysis of variance (ANOVA), which is a technique for breaking down the variance in a statistical sample into components that can be attributed to each factor affecting that sample. One major purpose of ANOVA is testing of the model. Model testing is important because of the serious consequences of erroneously concluding that a regression model is good when, in fact, it has little or no significance to the data. Model inadequacy often implies an error in the assumed relationships between the variables, poor data, or both. A validated regression model can be used for the following purposes:

1. Prediction/forecasting
2. Description
3. Control

Description of Regression Relationship

Sometimes, the desired result from a regression analysis is an equation describing the best fit to the data under investigation. The "least squares" line drawn through the data is the line of best fit. This line may be linear or curvilinear depending on the dispersion of the data. The linear situation exists in those cases where the slope of the regression equation is a constant. The nonconstant slope indicates curvilinear relationships. A plot of the data, called scatter plot, will usually indicate whether a linear or nonlinear model will be appropriate. The major problem with the nonlinear relationship is the necessity of assuming a functional relationship before accurately developing the model. Example of regression models (simple linear, multiple, and nonlinear) are presented as follows:

$$Y = \beta_0 + \beta_1 x + \varepsilon$$
$$Y = \beta_0 + \beta_1 x_1 + \beta_2 x_2 + \varepsilon$$
$$Y = \beta_0 + \beta_1 x_1^{\alpha 1} + \beta_2 x_2^{\alpha 2} + \varepsilon$$
$$Y = \beta_0 + \beta_1 x_1^{\alpha 1} + \beta_2 x_2^{\alpha 2} + \beta_{12} x_1^{\alpha 3} x_2^{\alpha 4} + \varepsilon$$

where
   Y is the dependent variable
   x_is are the independent variables
   β_is are the model parameters
   ε is the error term

The error terms are assumed to be independent and identically distributed normal random variables with mean of 0 and variance of σ^2.

Prediction

Another major use of regression analysis is prediction or forecasting. Prediction can be of two basic types: interpolation and extrapolation. Interpolation predicts values of the dependent variable over the range of the independent variable or variables. Extrapolation involves predictions outside the range of the independent variables. Extrapolation carries a risk in the sense that projections are made over a data range that is not included in the development of the regression model. There is some level of uncertainty about the nature of the relationships that may exist outside the study range. Interpolation can also create a problem when the values of the independent variables are widely spaced.

Control

Extreme care is needed in using regression for control. The difficulty lies in the assumption of a functional relationship when in fact none exists. Suppose, for example, that regression shows a relationship between chemical content in a product and noise level in the room. Suppose further that the real reason for this relationship is that the noise level increases as the machine speed increases and higher machine speed produces higher chemical content. It would be erroneous to assume a functional relationship between the noise level in the room and the chemical content in the product. If this relationship does exist, then changes in the noise level could control chemical content. In this case, the real functional relationship exists between machine speed and chemical content. It is often difficult to prove functional relationships outside a laboratory environment because many extraneous and intractable factors may have an influence on the dependent variable.

A simple example of the use of functional relationship for control can be seen in the following familiar equation of electrical circuits

$$I = \frac{V}{R}$$

where
  V is voltage
  I is electrical current
  R is the resistance

The current can be controlled by changes in either the voltage or the resistance or both. This particular equation, which has been experimentally validated, can be used as a control device.

PROCEDURE FOR REGRESSION ANALYSIS

Problem definition: Failure to properly define the scope of the problem could result in useless conclusions. Time can be saved throughout all phases of a regression study by knowing, as precisely as possible, the purpose of the required model. A proper definition of the problem will facilitate the selection of the appropriate variables to include in the study.

Selection of variables: Two very important factors in the selection of variables are ease of data collection and expense of data collection. Ease of data collection deals with the accessibility and the desired form of data. We must first determine if the data can be collected and, if so, how difficult the process will be. The economic question is of prime importance. How expensive will the data be to collect and compile into a useable form? If the expense cannot be justified, then the variable under consideration may necessarily be omitted from the selection process.

Test of significance of regression: After the selection and compilation of all possible relevant variables, the next step is a test for the significance of regression. The test should help avoid wasted effort on the use of an invalid model. The test for the significance of regression is a test to see if at least one of the variable coefficient(s) in the regression equation is statistically different from 0. A test indicating that none of the coefficients is significantly different from 0 implies that the best approximation of the data is a straight line through the data at the average value of the dependent variable regardless of the values of the independent variables. The significance level of the data is an indication of the probability of erroneously assuming model validity.

COEFFICIENT OF DETERMINATION

The coefficient of multiple determination, denoted by R^2, is used to judge the effectiveness of regression models containing multiple variables (multiple regression model). It indicates the proportion of the variation in the dependent variable explained by the model. The coefficient of multiple determination is defined as

$$R^2 = \frac{SSR}{SST}$$
$$= 1 - \frac{SSE}{SST}$$

where
SSR represents the sum of squares due to the regression model
SST represents the sum of squares total
SSE represents the sum of squares due to error

R^2 measures the proportionate reduction of total variation in the dependent variable accounted for by a specific set of independent variables. The coefficient of multiple determination, R^2, reduces to the *coefficient of simple determination*, r^2, when there is only one independent variable in the regression model. R^2 is equal to 0 when all the coefficients, b_k, in the model are 0. That is, no regression fit at all. R^2 is equal to 1 when all data points fall directly on the fitted response surface. Thus, we have

$$0.0 \leq R^2 \leq 1.0$$

The following points should be noted about regression modeling:

1. A large R^2 does not necessarily imply that the fitted model is a useful one. For example, observations may have been taken at only a few levels of the independent variables. In such a case, the fitted model may not be useful because most predictions would require extrapolation outside the region of observations. For example, for only two data points, the regression line passes perfectly through the two points and the R^2 value will be 1. In that case, despite the high R^2, there will be no useful prediction capability.
2. Adding more independent variables to a regression model can only increase R^2 and never reduce it. This is because the error sum of squares (SSE) can never become larger with more independent variables and the total sum of squares (SST) is always the same for a given set of responses.
3. Regression models developed under conditions where the number of data points is roughly equal to the number of variables will yield high values of R^2 even though the model may not be useful. For example, for only two data points, the regression line will pass perfectly through the two points and r^2 will be 1. Even though r^2 is 1, there will be no useful prediction.

The strategy for using R^2 to evaluate regression models should not entirely focus on maximizing the R^2 value. Rather, the intent should be to find the point where adding more independent variables is not worthwhile in terms of the overall effectiveness of the regression model. For most practical situations, R^2 values greater than 0.62 are considered acceptable.

Since R^2 can often be made larger by including a large number of independent variables, it is sometimes suggested that a modified measure be used which recognizes the number of independent variables in the model. This modified measure is referred to as adjusted coefficient of multiple determination, R_a^2. It is defined mathematically as

$$R_a^2 = 1 - \left(\frac{n-1}{n-p} \right) \frac{SSE}{SST}$$

where
  n is the number of observations used to fit the model
  p is the number of coefficients in the model (including the constant term)
  $p - 1$ is the number of independent variables in the model

R_a^2 may actually become smaller when another independent variable is introduced into the model. This is because the decrease in SSE may be more than offset by the loss of a degree of freedom in the denominator, $n - p$.

The *coefficient of multiple correlation* is defined as the positive square root of R^2. That is,

$$R = \sqrt{R^2}$$

Thus, the higher the value of R^2, the higher the correlation in the fitted model.

RESIDUAL ANALYSIS

A residual is the difference between the predicted value computed from the fitted model and the actual value from the data. The ith residual is defined as

$$e_i = Y_i - \hat{Y}_i$$

where
  Y_i is the actual value
  $\hat{Y}_i$ is the predicted value

The sum of squares of errors, *SSE*, and the mean square error, *MSE*, are computed as

$$SSE = \sum_i e_i^2$$

$$\sigma^2 \approx \frac{\sum_i e_i^2}{n-2} = MSE$$

where n is the number of data points. A plot of residuals versus predicted values of the dependent variable can be quite revealing. The plot for a good regression model will have a random pattern. A noticeable trend in the residual pattern indicates a problem with the model. Some possible reasons for an invalid regression model are as follows:

Insufficient data.
Important factors not included in model.
Inconsistency in data.
No functional relationship exists.

Graphical analysis of residuals is important for assessing the appropriateness of regression models. The different possible residual patterns are presented in Figure 3.1.

When we plot the residuals versus the independent variable, the result should appear ideally as shown in the first plot. The second plot shows a residual pattern indicating nonlinearity of the regression function. The third plot shows a pattern suggesting nonconstant variance (i.e., variation in σ^2). The fourth plot presents a

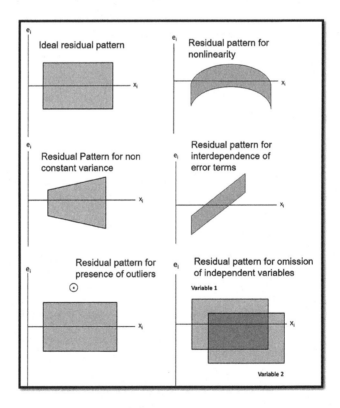

FIGURE 3.1   Six Residual Patterns.

residual pattern implying interdependence of the error terms. The fifth plot shows a pattern depicting the presence of outliers. The sixth and last plot represents a pattern suggesting omission of independent variables.

TIME SERIES ANALYSIS

Time series analysis is a technique that attempts to predict the future by using historical data. The basic principle of time series analysis is that the sequence of observations is based on jointly distributed random variables. The time series observations denoted by Z_1, Z_2, ..., Z_T are assumed to be drawn from some joint probability density function (pdf) of the form

$$f_{1,\ldots,T}\left(Z_{1,\ldots,T}\right)$$

The objective of time series analysis is to use the joint density to make probability inferences about future observations. The concept of stationarity implies that the distribution of the time series is invariant with regard to any time displacement. That is,

$$f\left(Z_t,\ldots,Z_{t+k}\right) = f\left(Z_{t+m},\ldots,Z_{t+m+k}\right)$$

where
t is any point in time
k and m are any pair of positive integers

A stationary time series process has a constant variance and remains stable around a constant mean with respect to time reference. Thus,

$$E\left(Z_t\right) = E\left(Z_{t+m}\right)$$
$$V\left(Z_t\right) = V\left(Z_{t+m}\right)$$
$$\text{cov}\left(Z_t,Z_{t+1}\right) = \text{cov}\left(Z_{t+k},Z_{t+k+1}\right)$$

Nonstationarity in a time series may be recognized in a plot of the series. A widely scattered plot with no tendency for a particular value is an indication of nonstationarity.

STATIONARITY AND DATA TRANSFORMATION

In some cases where nonstationarity exists, some form of data transformation may be used to achieve stationarity. For most time series data, the usual transformation tried is differencing. Differencing involves the creation of a new series by taking differences between successive periods of the original series. For example, first regular differences are obtained by

$$w_t = Z_t - Z_{t-1}$$

To develop a time series forecasting model, it is necessary to describe the relationship between a current observation and previous observations. Such relationships are described by the sample autocorrelation function defined as shown in the following formula:

$$r_j = \frac{\sum_{t=1}^{T-j}\left(Z_t - \bar{Z}\right)\left(Z_{t+j} - \bar{Z}\right)}{\sum_{t=1}^{T}\left(Z_t - \bar{Z}\right)^2}, \quad j = 0,1,\ldots,T-1$$

where

T is the number of observations
Z_t is the observation for time t
$\bar{Z}$ is the sample mean of the series
j is the number of periods separating pairs of observations
r_j is the sample estimate of the theoretical correlation coefficient

The coefficient of correlation between two variables Y_1 and Y_2 is defined as

$$\rho_{12} = \frac{\sigma_{12}}{\sigma_1 \sigma_2}$$

where

σ_1 and σ_2 are the standard deviations of Y_1 and Y_2, respectively
σ_{12} is the covariance between Y_1 and Y_2

The standard deviations are the positive square roots of the variances defined as

$$\sigma_1^2 = E\left[(^{Y_1}\right]$$
$$\sigma_2^2 = E\left[(^{Y_2}\right]$$

The covariance, σ_{12}, is defined as

$$\sigma_{12} = E\left[\left(Y_1 - \mu_1\right)\left(Y_2 - \mu_2\right)\right]$$

which will be 0 if Y_1 and Y_2 are independent. Thus, when $\sigma_{12} = 0$, we also have $\rho_{12} = 0$. If Y_1 and Y_2 are positively related, then σ_{12} and ρ_{12} are both positive. If Y_1 and Y_2 are negatively related, then σ_{12} and ρ_{12} are both negative. The correlation coefficient is a real number between -1 and $+1$:

$$1.0 \le \rho_j \le +1.0$$

Time series modeling procedure involves the development of a discrete linear stochastic process in which each observation, Z_t, may be expressed as

$$Z_t = \mu + \mu_t + \Psi_1 u_{t-1} + \Psi_2 u_{t-2} + \cdots$$

where
  μ is the mean of the process
  Ψ_is are model parameters, which are functions of the autocorrelations

Note that this is an infinite sum indicating that the current observation at time t can be expressed in terms of all previous observations from the past. In a practical sense, some of the coefficients will be 0 after some finite point q in the past. The u_ts form the sequence of independently and identically distributed random disturbances with mean 0 and variance sigma sub u^2. The expected value of the series is obtained by

$$\begin{aligned} E(Z_t) &= \mu + E(u_t + \Psi_1 u_{t-1} + \Psi_2 u_{t-2} + \cdots) \\ &= \mu + E(u_t)[1 + \Psi_1 + \Psi_2 + \cdots] \end{aligned}$$

Stationarity of the time series requires that the expected value be stable. That is, the infinite sum of the coefficients should be convergent:

$$\sum_{i=0}^{\infty} \Psi_i = c$$

where
  $\Psi_0 = 1$
  c is a constant

The theoretical variance of the process, denoted by γ_0, can be derived as

where σ_u^2 represents the variance of the u_ts. The theoretical covariance between Z_t and Z_{t+j}, denoted by γ_j, can be derived in a similar manner to obtain

$$\begin{aligned} \gamma_j &= E\{[Z_t - E(Z_t)][Z_{t+j} - E(Z_{t+j})]\} \\ &= \sigma_u^2(\Psi_j + \Psi_1 \Psi_{j+1} + \Psi_2 \Psi_{j+2} + \cdots) \\ &= \sigma_u^2 \sum_{i=0}^{\infty} \Psi_{i+j} \Psi \end{aligned}$$

The expectation of the cross-product terms is zero since the u_t's are independently distributed.

Sample estimates of the variances and covariances are obtained by

$$c_j = \frac{1}{T}\sum_{t=1}^{T-j}\left(Z_t - \bar{Z}\right)\left(Z_{t+j} - \bar{Z}\right), \quad j = 0,1,2,\ldots$$

The theoretical autocorrelations are obtained by dividing each of the autocovariances, γ_j, by γ_0. Thus, we have

$$\rho_j = \frac{\gamma_j}{\gamma_0}, \quad j = 0,1,2,\ldots$$

and the sample autocorrelations are obtained by

$$r_j = \frac{c_j}{c_0}, \quad j = 0,1,2,\ldots$$

Moving Average Processes

If it can be assumed that $\Psi_i = 0$ for some $i > q$, where q is an integer, then our time series model can be represented as

$$z_t = \mu + u_t + \Psi_1 u_{t-1} + \Psi_2 u_{t-2} + \cdots + \Psi_q u_{t-q}$$

which is referred to as a moving average process of order q, usually denoted as $MA(q)$. For notational convenience, we denote the truncated series as presented in the following:

$$Z_t = \mu + u_t - \theta_1 u_{t-1} - \theta_2 u_{t-2} - \cdots - \theta_q u_{t-q}$$

where $\theta_0 = 1$. Any $MA(q)$ process is stationary since the condition of convergence for the Ψ_is becomes

$$\left(1 + \Psi_1 + \Psi_2 + \cdots\right) = \left(1 - \theta_1 - \theta_2 - \cdots - \theta_q\right)$$

$$= 1 - \sum_{i=0}^{q}\theta_i$$

which converges since q is finite. The variance of the process now reduces to

$$\gamma_0 = \sigma_u^2 \sum_{i=0}^{q}\theta_i$$

We now have the autocovariances and autocorrelations defined, respectively, as shown in the following:

$$\gamma_j = \sigma_u^2 \left(-\theta_j + \theta_1\theta_{j+1} + \cdots + \theta_{q-j}\theta_q \right), \quad j = 1, \ldots, q$$

where $\gamma_j = 0$ for $j > q$

$$\rho_j = \frac{\left(-\theta_j + \theta_1\theta_{j+1} + \cdots + \theta_{q-j}\theta_q \right)}{\left(1 + \theta_1^2 + \cdots + \theta_q^2 \right)}, \quad j = 1, \ldots, q$$

where $\rho_j = 0$ for $j > q$.

Autoregressive Processes

In the preceding section, the time series, Z_t, is expressed in terms of the current disturbance, ut, and past disturbances, $ut - i$. An alternative is to express Z_t, in terms of the current and past observations, Z_{t-i}. This is achieved by rewriting the time series expression as

$$u_t = Z_t - \mu - \Psi_1 u_{t-1} - \Psi_2 u_{t-2} - \cdots$$
$$u_{t-1} = Z_{t-1} - \mu - \Psi_1 u_{t-2} - \Psi_2 u_{t-3} - \cdots$$
$$u_{t-2} = Z_{t-2} - \mu - \Psi_1 u_{t-3} - \Psi_2 u_{t-4} - \cdots$$

Successive back substitutions for the u_{t-i}s yields the following:

$$u_t = \pi_1 Z_{t-1} - \pi_2 Z_{t-2} - \cdots - \delta$$

where π_is and δ are model parameters and are functions of Ψ_is and μ. We can then rewrite the model as

$$Z_t = \pi_1 Z_{t-1} + \pi_2 Z_{t-2} + \cdots + \pi_p Z_{t-p} + \delta + u_t$$

which is referred to as an *autoregressive process of order p*, usually denoted as *AR(p)*. For notational convenience, we denote the autoregressive process as shown in the following:

$$Z_t = \varphi_1 Z_{t-1} + \varphi_2 Z_{t-2} + \cdots + \varphi_p Z_{t-p} + \delta + u$$

Thus, AR processes are equivalent to MA processes of infinite order. Stationarity of AR processes is confirmed if the roots of the following characteristic equation lie outside the unit circle in the complex plane:

$$\left(1 - \varphi_1 x - \varphi_2 x^2 - \cdots - \varphi_p x^p \right) = 0$$

where x is a dummy algebraic symbol. If the process is stationary, then we should have

$$E(Z_t) = \varphi_1 E(Z_{t-1}) + \varphi_2 E(Z_{t-2}) + \cdots + \varphi_p E(Z_{t-p}) + \delta + E(u_t)$$
$$= \varphi_1 E(Z_t) + \varphi_2 E(Z_t) + \cdots + \varphi_p E(Z_t) + \delta$$
$$= E(Z_t)(\varphi_1 + \varphi_2 + \cdots + \varphi_p) + \delta$$

which yields

$$E(Z_t) = \frac{\delta}{(1 - \varphi_1 - \varphi_2 - \cdots - \varphi_p)}$$

Denoting the deviation of the process from its mean by Z_t^d, the following is obtained:

$$Z_t^d = Z_t - E(Z_t) = Z_t - \frac{\delta}{(1 - \varphi_1 - \varphi_2 - \cdots - \varphi_p)}$$

$$Z_{t-1}^d - Z_{t-1} - \frac{\delta}{(1 - \varphi_1 - \varphi_2 - \cdots - \varphi_p)}$$

Rewriting the previous expression yields

$$Z_{t-1} = Z_{t-1}^d + \frac{\delta}{(1 - \varphi_1 - \varphi_2 - \cdots - \varphi_p)}$$

$$\cdots$$

$$\cdots$$

$$Z_{t-k} = Z_{t-k}^d + \frac{\delta}{(1 - \varphi_1 - \varphi_2 - \cdots - \varphi_p)}$$

If we substitute the $AR(p)$ expression into the expression for Z_t^d, we obtain

$$Z_t^d = \varphi_1 Z_{t-1} + \varphi_2 Z_{t-2} + \cdots + \varphi_p Z_{t-p} + \delta + u_t - \frac{\delta}{(1 - \varphi_1 - \varphi_2 - \cdots - \varphi_p)}$$

Successive back substitutions of $Z_t - {}_j$ into the preceding expression yields

$$Z_t^d = \varphi_1 Z_{t-1}^d + \varphi_2 Z_{t-2}^d + \cdots + \varphi_p Z_{t-p}^d + u_t$$

Thus, the deviation series follows the same AR process without a constant term. The tools for identifying and constructing time series models are the sample autocorrelations, r_j. For the model identification procedure, a visual assessment of the plot of r_j against j, called the sample correlogram, is used. Figure 3.2 presents examples of *sample correlograms* and the corresponding time series models.

A wide variety of sample correlogram patterns can be encountered in time series analysis. It is the responsibility of the analyst to choose an appropriate model to fit the prevailing time series data. Several statistical computer programs are available for performing time series analysis.

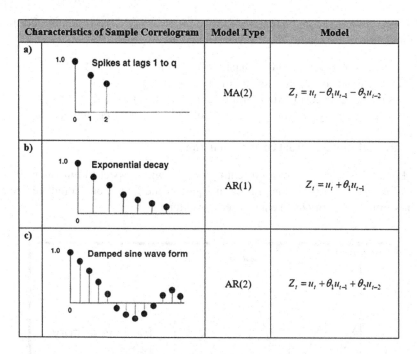

Characteristics of Sample Correlogram	Model Type	Model
a) Spikes at lags 1 to q	MA(2)	$Z_t = u_t - \theta_1 u_{t-1} - \theta_2 u_{t-2}$
b) Exponential decay	AR(1)	$Z_t = u_t + \theta_1 u_{t-1}$
c) Damped sine wave form	AR(2)	$Z_t = u_t + \theta_1 u_{t-1} + \theta_2 u_{t-2}$

FIGURE 3.2   Characteristics of Sample Correlogram.

INVENTORY MANAGEMENT MODELS

Inventoried items are an important component of any supply chain, even in non-manufacturing operations. Any resource can be viewed as an "inventoried" item for the purpose of managing the supply chain. Consequently, inventory management strategies are essential for comprehensive supply chain planning and control. Tracking activities is analogous to tracking inventory items. The important aspects of inventory management for supply systems management are:

1. Ability to satisfy work demands promptly by supplying materials from stock
2. Availability of bulk rates for purchases and shipping
3. Possibility of maintaining more stable and level resource or workforce

Some of the basic and classical inventory control techniques are discussed in the following.

Economic Order Quantity Model

The economic order quantity (*EOQ*) model determines the optimal order quantity based on purchase cost, inventory carrying cost, demand rate, and ordering cost.

The objective is to minimize the total relevant costs (*TRC*) of inventory. For the formulation of the model, the following notations are used:

Q is the replenishment order quantity (in units)
A is the fixed cost of placing an ordering
v is the variable cost per unit of the item to be inventoried
r is the inventory carrying charge per dollar of inventory per unit time
D is the demand rate of the item
TRC is the total relevant costs per unit time

Figure 3.3 shows the basic inventory pattern with respect to time as well as some inventory cost patterns. In the first part of the figure, one complete cycle starts from a level of Q and ends at zero inventory.

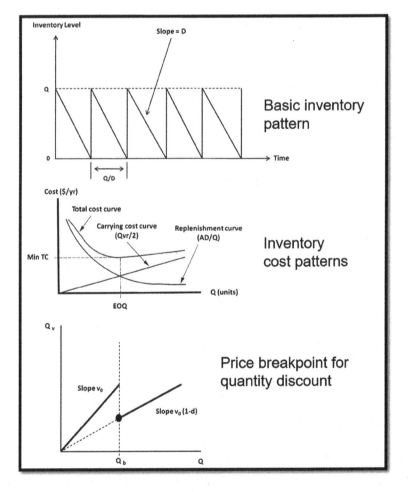

FIGURE 3.3   Inventory Pattern and Cost Plots.

The second part of the figure shows the costs as functions of replenishment quantity. The *TRC* for order quantity Q is given by the expression as follows:

$$TRC(Q) = \frac{Qvr}{2} + \frac{AD}{Q}$$

When the *TRC(Q)* function is optimized with respect to Q, we obtain the expression for the *EOQ*:

$$EOQ = \sqrt{\frac{2AD}{vr}}$$

which represents the minimum *TRC* of inventory. The previous formulation assumes that the cost per unit is constant regardless of the order quantity. In some cases, quantity discounts may be applicable to the inventory item. The formulation for quantity discount situation is presented in the following.

QUANTITY DISCOUNT

A quantity discount may be available if the order quantity exceeds a certain level. This is referred to as the single breakpoint discount. The third part of Figure 3.3 presents the price breakpoint for quantity discount.

The unit cost is represented in the following equation:

$$v = \begin{cases} v_0, & 0 \leq Q < Q_b \\ v_0(1-d), & Q_b \leq Q \end{cases}$$

where
v_0 is the basic unit cost without discount
d is the discount (in decimals)
d is applied to all units when $Q \leq Q_b$
Q_b is the breakpoint

CALCULATION OF TOTAL RELEVANT COST

For $0 \leq Q < Q_b$, we obtain

$$TRC(Q) = \left(\frac{Q}{2}\right)v_0 r + \left(\frac{A}{Q}\right)D + Dv_0$$

For $Q_b \leq Q$, we have

$$TRC(Q)_{discount} = \left(\frac{Q}{2}\right)v_0(1-d)r + \left(\frac{A}{Q}\right)D + Dv_0(1-d)$$

Note that for any given value of Q, $TRC(Q)_{discount} < TRC(Q)$. Therefore, if the lowest point on the $TRC(Q)_{discount}$ curve corresponds to a value of $Q^* > Q_b$ (i.e., Q is valid), then set $Q_{opt} = Q^*$.

EVALUATION OF THE DISCOUNT OPTION

The trade-off between extra carrying costs and the reduction in replenishment costs should be evaluated to see if the discount option is cost justified. Reduction in replenishment cost is composed of:

1. Reduction in unit value
2. Fewer replenishments per unit time
   Case a: If reduction in acquisition costs > extra carrying costs, then set $Q_{opt} = Q_b$.
   Case b: If reduction in acquisition costs < extra carrying costs, then set $Q_{opt} = EOQ$ with no discount.
   Case c: If Q_b is relatively small, then set $Q_{opt} = EOQ$ with discount. The optimal order quantity, Q_{opt}, can be found as follows:
   Step 1: Compute EOQ when d is applicable:

$$EOQ(\text{discount}) = \sqrt{\frac{2AD}{v_0(1-d)r}}$$

   Step 2: Compare $EOQ(d)$ with Q_b:
   If $EOQ(d) \geq Q_b$, set $Q_{opt} = EOQ(d)$
   If $EOQ(d) < Q_b$, go to *Step 3*
   Step 3: Evaluate TRC for EOQ and Q_b:
   Suppose $d = 0.02$ and $Q_b = 100$ for the three items shown below:
   Item 1: D (Units/yr) = 416; v_0($/Unit) = 14.20; (A) ($) = 1.50; r ($/$/year) = 0.24
   Item 2: D (Units/yr) = 104; v_0($/Unit) = 3.10; (A) ($) = 1.50; r ($/$/year) = 0.24
   Item 3: D (Units/yr) = 4,160; v_0($/Unit) = 2.40; (A) ($) = 1.50; r ($/$/year) = 0.24

Item 1:
Step 1: EOQ(discount) = 19 units < 100 units
Step 2: EOQ(discount) < Q_b, go to *Step 3*
Step 3: Calculate TRC values, using the following equation:

$$TRC(Q) = \left(\frac{Q}{2}\right)v_0 r + \left(\frac{A}{Q}\right)D + Dv_0$$

Since $TRC(EOQ) > TRC(Q_b)$, set $Q_{opt} = 100$ units.

Item 2:

Step 1: EOQ(discount) = 21 units < 100 units

Step 2: EOQ(discount) < Qb, go to Step 3

Step 3: TRC values

$$TRC(EOQ) = \sqrt{2(1.50)(104)(3.10)(0.24)} + 104(3.10)$$
$$= \$337.64 \, / \, year$$

$$TRC(Q_b) = \frac{100(3.10)(0.98)(0.24)}{2} + \frac{(1.50)(104)}{100} + 104(3.10)(0.98)$$
$$= \$353.97 \, / \, year$$

$$TRC(EOQ) < TRC(Q_b), \quad \text{set} \, Q_{opt} = EOQ(\text{without discount}):$$
$$EOQ = \sqrt{\frac{2(1.50)(104)}{3.10(0.24)}}$$
$$= 20 \text{units}$$

Item 3:

Step 1: Compute EOQ(discount)

$$EOQ(\text{discount}) = \sqrt{\frac{2(1.50)(4160)}{2.40(0.98)(0.24)}}$$
$$= 149 > 100 \text{units}$$

Step 2: EOQ(discount) > Q_b. Set Q_{opt} = 149 units.

SENSITIVITY ANALYSIS

Sensitivity analysis involves a determination of the changes in the values of a parameter that will lead to a change in a dependent variable. It is a process for determining how wrong a decision will be if assumptions on which the decision is based prove to be incorrect. For example, a "decision" may be dependent on the changes in the values of a particular parameter and inventory cost may be the parameter on which the decision depends. Cost itself may depend on the values of other parameters as shown in the following:

$$Sub - parameter \rightarrow Main \, parameter \rightarrow Decision$$

It is of interest to determine what changes in parameter values can lead to changes in a decision. With respect to inventory management, we may be interested in the cost impact of deviation of actual order quantity from the *EOQ*. The sensitivity of cost to departures from *EOQ* is analyzed as presented in the following:

Let *p* represent the level of change from *EOQ*:

$$|p| \leq 1.0$$
$$Q' = (1 - p)EOQ$$

Percentage cost penalty (PCP) is defined as

$$PCP = \frac{TRC(Q') - TRC(EOQ)}{TRC(EOQ)}(100)$$
$$= 50\left(\frac{p^2}{1+p}\right)$$

A plot of the sensitivity with respect to the respective *PCP* may be developed to provide a visual assessment. It can be seen that the cost is not very sensitive to minor departures from *EOQ*. We can conclude that changes within 10% of *EOQ* will not significantly affect the *TRC*. There are several inventory control algorithms available in the literature. Two examples are the Wagner–Whitin (W–W) algorithm and the Silver-Meal heuristic (Badiru, 2019). The W–W algorithm is an approach to deterministic inventory model. It is based on dynamic programming technique. The Silver-Meal heuristic is a simple inventory control technique that is recommended for items that have significantly variable demand pattern. Its objective is to minimize *TRC* per unit time for the duration of the replenishment quantity:

$$TRC = A + \text{Carrying costs}$$

where *A* is the cost of placing an order.

MODELING FOR SEASONAL PATTERN

There are numerous applications of statistical distributions to practical, real-world problems. Several distributions have been developed and successfully utilized for a large variety of problems. Despite the large number of distributions available, it is often confusing to determine which distribution is applicable to which real-world random variable. In many applications, the choices are limited to a few familiar distributions due either to a lack of better knowledge or computational ease. Such familiar distributions include the *normal, exponential,* and *uniform* distributions. To accommodate cases where the familiar distributions do not adequately represent the variable of interest, some special purpose distributions have been developed. Examples of such special distributions are *Pareto, Rayleigh, lognormal,* and *Cauchy* distributions. Consider the *cyclic probability density function* (also called *periodic distribution* or *seasonal distribution)* for special cases involving random variables that are governed by cyclic processes. These special cases include seasonal inventory control and time series processes. Instabilities in a system can be caused by erratic materials supply, cyclic inventory, and seasonal work patterns. A time series is a collection of observations that are drawn from a periodic or cyclic process. Examples of operating time series include monthly cost, quarterly revenue, and seasonal energy consumption. The standard cyclic pdf is a continuous waveform periodic function defined on the interval $[0, 2\pi)]$. The standard function is transformed into a general cyclic function defined over any time series interval $[a, b]$. One possible application of the cyclic distribution is the estimation of time to failure for components or equipment that undergoes periodic maintenance. Another application may be in the statistical analysis of the peak levels in a time series process.

The trigonometric sine function (Badiru, 2019) provides the basis for the cyclic distribution. The basic sine function is given by

$$y = \sin\theta, \quad -\infty < \theta < \infty$$

where θ is in radians. The basic sine function and its variations are shown in Figure 3.4.

The function is cyclic with period 2π. The sine function has its maxima and minima (1 and −1, respectively) at the points $\theta = \pm n\pi/2$, where n is an odd positive integer. The function also satisfies the relationship

$$\sin(\theta + 2n\pi) = \sin\theta$$

for any integer n. In general, any function $f(x)$ satisfying the relationship

$$f(x + nT) = f(x)$$

is said to be cyclic with period T, where T is a positive constant and n is an integer. A graph of $f(x)$ truncated to an interval $[(a, a + T)$ or $(a, a + T)]$ is called *one cycle of the function.*

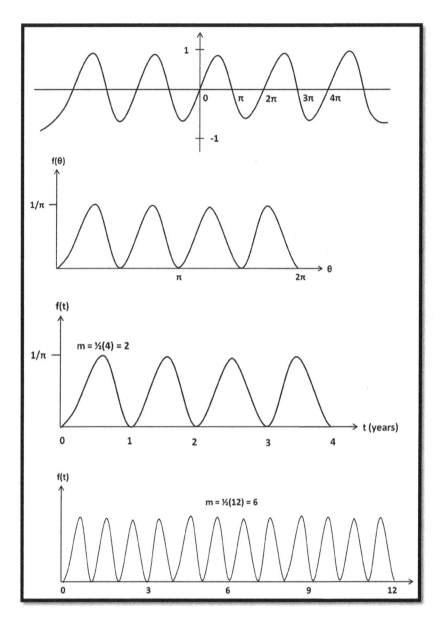

FIGURE 3.4   Basic Sine Function and Its Variations.

Standard Cyclic PDF

The cyclic pdf is defined as

$$f(x) = \begin{cases} \dfrac{1}{\pi}\sin^2 mx & 0 \le x \le 2\pi, \quad m = 1,2,3,\ldots \\ 0 & \textit{otherwise} \end{cases}$$

where
  x is in radians
  m (positive integer) is the number of peaks (or modes) associated with the
  function on the interval $[0, \pi]$, which is half of the period

Thus, $f(x)$ represents a multimodal distribution. To be a legitimate pdf, $f(x)$ must satisfy the following conditions:

1. $f(x) \ge 0$ for all x in $[0, 2\pi]$.

2. $\displaystyle\int_0^{2\pi} f(x)\,dx = 1$.

3. $P(x_1 < X < x_2) = \displaystyle\int_{x_1}^{x_2} f(x)\,dx$.

The first condition can be seen to be true by inspection. Each value of x yields a probability value equal to or greater than 0. The second condition can be verified by examining the definite integral

$$\int_0^{2\pi} f(x)\,dx = \frac{1}{\pi} \int_0^{2\pi} \sin^2(mx)\,dx$$

$$= \frac{1}{\pi}\left[\frac{x}{2} - \frac{\sin(2mx)}{4m} \right]_0^{2\pi}$$

$$= \frac{1}{\pi}\left[\pi - \frac{\sin(4mx)}{4m} \right]$$

$$= 1$$

The third condition is easily verified by computing the probability that X will fall between x_1 and x_2 as the area under the curve for $f(x)$ from x_1 to x_2.

Expected Value

The expected value of X is given by

$$E[X] = \int_0^{2\pi} x f(x)\,dx$$

$$= \frac{1}{\pi} \int_0^{2\pi} x \sin^2(mx)\,dx$$

$$= \frac{1}{\pi}\left[\frac{x^2}{4} - \frac{x\sin(2mx)}{4m} - \frac{\cos(2mx)}{8m^2}\right]_0^{2\pi}$$

$$= \pi$$

Variance

The variance of X is calculated from the theoretical definition of variance, which makes use of the definite integral from 0 to 2π. This is given by

Cumulative Distribution Function

The cumulative distribution function of $f(x)$ is given by

$$F(x) = \int_0^x f(r)\,dr$$

$$= \int_0^x \frac{1}{\pi}\sin^2(mr)\,dr$$

$$= \frac{2mx - \sin(2mx)}{4m\pi}$$

General Cyclic PDF

For practical applications, a general form of the cyclic pdf defined over any real interval $[a, b]$ will be of more interest. This general form is given by:

$$f(x) = \begin{cases} \dfrac{1}{\pi}\sin^2 mr & 0 \le r \le 2\pi, \quad m = 1,2,3,\dots \\ 0 & \text{otherwise} \end{cases}$$

where

m is half of the number of peaks expected over the interval $[a, b]$

r is the radian equivalent of the real number x in the interval $[a, b]$

The transformations from real units to radians and vice versa are accomplished by the expression as follows:

$$r = 2\pi \frac{x-a}{b-a}, \quad 0 \le r \le 2\pi, \quad a \le x \le b$$

$$x = a + (r)\frac{b-a}{2\pi}, \quad 0 \le r \le 2\pi, \quad a \le x \le b$$

Using the earlier transformation relationships and the expressions derived previously for the standard cyclic random variable, the mean and variance of the general cyclic random variable are derived to be

$$E[X] = a + \frac{b-a}{2\pi}(E[r])$$

$$= a + \frac{b-a}{2\pi}(\pi)$$

$$= \frac{a+b}{2}$$

$$V[X] = 0 + \left(\frac{b-a}{2\pi}\right)^2 V[r]$$

$$= \frac{(b-a)^2}{4\pi^2}\left(\frac{\pi^2}{3}\right)$$

$$= \frac{(b-a)^2}{12}$$

Note that these expressions are identical to the expressions for the mean and variance of the uniform distribution. Badiru (2019) presents application examples of the sine function for seasonal inventory patterns.

CONCLUSION

Forecasting is a basic requirement for a responsive and adaptive supply chain. Coupled with inventory modeling, practical data analytics can be performed as the basis for supply chain decisions. This chapter has presented a basic analysis of forecasting and inventory modeling in the context of supply chain systems.

REFERENCES

Badiru, A. B. (2019), *Project Management: Systems, Principles, and Applications*, 2nd Edition, Taylor & Francis Group/CRC Press, Boca Raton, FL.

Badiru, A. B., P. S. Pulat and M. Kang 1993, "DDM: decision support system for hierarchical dynamic decision making," *Decision Support Systems*, Vol. 10, No. 1, pp. 1–18.

Hogan, Dakotah, John Elshaw, Clay Koschnick, Jonathan Ritschel, Adedeji Badiru, and Shawn Valentine (2020), "Cost estimating using a new learning curve theory for non-constant production rates," *Forecasting*, Vol. 2, No. 4, October 2020, pp. 429–451.

4 Modeling for Supply Chain Optimization

INTRODUCTION TO SCHEDULE OPTIMIZATION

Schedule optimization is often the major focus in project management. While heuristic scheduling is very simple to implement, it does have some limitations. The limitations of heuristic scheduling include subjectivity, arbitrariness, and simplistic assumptions. In addition, heuristic scheduling does not handle uncertainty very well. On the other hand, mathematical scheduling is difficult to apply to practical problems. However, the increasing access to low-cost high-speed computers has facilitated increased use of mathematical scheduling approaches that yield optimal project schedules. The advantages of mathematical scheduling include the following facts:

It provides optimal solutions.
It can be formulated to include realistic factors influencing a project.
Its formulation can be validated.
It has proven solution methodologies.

With the increasing availability of personal computers and software tools, there is very little need to solve optimization problems by hand nowadays. Computerized algorithms are now available to solve almost any kind of optimization problem. What is more important for the project analyst is to be aware of the optimization models available, the solution techniques available, and how to develop models for specific project optimization problems. It is crucial to know which model is appropriate for which problem and to know how to implement optimized solutions in practical settings. The presentation in this chapter concentrates on the processes for developing models for project optimization as presented by Badiru and Pulat (1995).

GENERAL SUPPLY CHAIN SCHEDULING FORMULATION

Several mathematical models can be developed for project scheduling problems, depending on the specific objective of interest and the prevailing constraints. One general formulation is

$$\text{Minimize}: \left\{ \max_{\forall i} \left\{ s_i + d_i \right\} \right\}$$

$$\text{Subject to}: s_i \geq s_j + d_j \quad \text{for all } i; j \in P_i$$

$$R_k \geq \sum_{i \in A_t} r_{ik} \quad \text{for all } t; \text{for all } k$$

$$s_i \geq 0 \quad \text{for all } i$$

$$r_{ik} \geq 0 \quad \text{for all } i; \text{for all } k$$

where
s_i is the start time of activity i
d_i is the duration of activity i
P_i is the set of activities which must precede activity i
R_k is the availability level of resource type k over the project horizon
A_t is the set of activities ongoing at time t
r_{ik} is the number of units of resource type k required by activity i

The objective of the aforementioned model is to minimize the completion time of the last activity in the project. Since the completion time of the last activity determines the project duration, the project duration is indirectly minimized. The first constraint set ensures that all predecessors of activity i are completed before activity i may start. The second constraint set ensures that resource allocation does not exceed resource availability. The general model may be modified or extended to consider other project parameters. Examples of other factors that may be incorporated into the scheduling formulation include cost, project deadline, activity contingency, mutual exclusivity of activities, activity crashing requirements, and activity subdivision.

An *objective function* is a mathematical representation of the goal of an organization. It is stated in terms of maximizing or minimizing some quantity of interest. In a project environment, the objective function may involve any of the following:

Minimize project duration.
Minimize project cost.
Minimize number of late jobs.
Minimize idle resource time.
Maximize project revenue.
Maximize net present worth.

LINEAR PROGRAMMING FORMULATION

Although many optimization models have emerged over the past decades, the linear programming (LP) formulation remains the seminal model, from which many enhancements have been developed. If the formulation and framework of the LP is understood, the whole concept of mathematical optimization becomes easier to imbibe.

The LP is a mathematical technique for maximizing or minimizing some quantity, such as profit, cost, or time to complete a project. It is one of the most widely used quantitative techniques. It is a mathematical technique for finding the optimum solution to a linear objective function of two or more quantitative decision variables subject to a set of linear constraints. The technique is applicable to a wide range of decision-making problems. Its wide applicability is due to the fact that its formulation is not tied to any particular class of problems, as the CPM and PERT techniques are. Numerous research and application studies of the LP are available in the literature.

The objective of a LP model is to optimize an objective function by finding values for a set of decision variables subject to a set of constraints. We can define the optimization problem mathematically as

$$z = c_1 \sum_{j=1}^{n1} x_{1j} + c_2 \sum_{j=1}^{n2} x_{2j} \ldots\ldots + c_k \sum_{j=1}^{N} x_{kj}$$

where
Optimize is replaced by *maximize* or *minimize* depending on the objective
z is the value of the objective function for specified values of the decision variables
$x_1, x_2, \ldots, x_n$ are the n decision variables.
$c_1, c_2, \ldots, c_n$ are the objective function coefficients.
k is the number of variables in the decision problem.
j and n are the counting indices for each decision variable, if applicable.

The word *programming* in the LP does not refer to computer programming, as some people think. Rather, it refers to choosing a *program of action*. The word *linear* refers to the *linear relationships* among the variables in the model. The characteristics of the LP formulation are explained next.

Quantitative decision variables. A decision variable is a factor that can be manipulated by the decision-maker. Examples are number of resource units assigned to a task, number of product types in a product mix, and number of units of a product to produce. Each decision variable must be defined numerically in some unit of measurement.

Linear objective function. The objective function relates to the measure of performance to be minimized or maximized. There is a linear relationship among the variables that make up the objective function. The coefficient of each variable in the objective function indicates its per unit contribution (positive or negative) toward the value of the objective function.

Linear constraints. Every decision problem is subject to some specific limitations or constraints. The constraints specify the restrictions on how the decision-maker may manipulate the decision variables. Examples of decision constraints are capacity limitations, maximum number of resource units available, demand and supply requirements, and number of work hours per day. The relationships among the variables in constraint must be expressed as linear functions represented as equations or inequalities.

Nonnegativity constraint. The nonnegativity constraint is common to all the LP problems. This requires that all decision variables are restricted to nonnegative values.

The general procedure for using a LP model to solve a decision problem involves an LP formulation of the problem and a selection of a solution approach. The procedure is summarized as follows:

1. Determine the decision variables in the problem.
2. Determine the objective of the problem.
3. Formulate the objective function as an algebraic expression.
4. Determine the real-world restrictions on the problem scenario.
5. Write each of the restrictions as an algebraic constraint. Make sure that units match throughout the constraints. Otherwise, the terms cannot be added.
6. Select a solution approach. The *graphical method* and the *simplex technique* are the two most popular approaches. The graphical method is easy to apply when the LP model contains just two decision variables. Several commercial software packages are available for solving the LP and related formulations.

An important aspect of using the LP models is the interpretation of the results to make decisions. An LP solution that is optimal analytically may not be practical in a real-world decision scenario. The decision-maker must incorporate his or her own subjective judgment when implementing the LP solutions. Final decisions are often based on a combination of quantitative and qualitative factors. The examples presented in this chapter illustrate the application of optimization models to project planning and scheduling problems.

ACTIVITY PLANNING FORMULATION

Activity planning is a major function in any supply chain. The LP can be used to determine the optimal allocation of time and resources to the activities in a project. Suppose a program planner is faced with the problem of planning a 5-day development program for a group of managers in a manufacturing organization. The program includes some combination of four activities: a seminar, laboratory work, case studies, and management games. It is estimated that each day spent on an activity will result in productivity improvement for the organization. The productivity improvement will generate annual cost savings as shown in Table 4.1.

TABLE 4.1
Data for Activity Planning Problem

Activity	Cost savings ($/year)	% Active	% Passive	Cost ($/day)
Seminar	3,200,000	10	90	400
Laboratory work	2,000,000	40	60	200
Case studies	400,000	100	0	75
Management games	2,000,000	60	40	100

The program will last 5 days and there is no time lost between activities. In order to balance the program, the planner must make sure that not more than 3 days are spent on active or passive elements of the program. The active and passive percentages of each activity are also shown in the table. The company wishes to spend at least 0.5 day on each of the four activities. A total budget of $1500 is available. The cost of each activity is shown in the tabulated data.

The program planner must determine how many days to spend on each of the four activities. The following variables are defined for the problem:

x_1 represents number of days spent on a seminar
x_2 represents number of days of laboratory work
x_3 represents number of days for case studies
x_4 represents number of days with management games

The objective is to maximize the estimated annual cost savings. That is,

$$\text{Maximize}: f = 3200x_1 + 2000x_2 + 400x_3 + 2000x_4$$

Subject to the following constraints:

1. The program lasts exactly 5 days.

$$x_1 + x_2 + x_3 + x_4 = 5$$

2. Not more than 3 days can be spent on active elements:

$$0.10x_1 + 0.40x_2 + x_3 + 0.60x_4 \leq 3$$

3. Not more than 3 days can be spent on passive elements:

$$0.90x_1 + 0.60x_2 + 0.40x_4 \leq 3$$

TABLE 4.2
The LP Solution to the Activity Planning Example

Activity	Cost Savings ($/year)	Number of Days	Annual Cost Savings ($)
Seminar	3,200,000	2.20	7,040,000
Laboratory work	2,000,000	0.50	1,000,000
Case studies	400,000	0.50	200,000
Management games	2,000,000	1.80	3,600,000
Total		5	11,840,000

4. At least 0.5 day most be spent on each of the four activities:

$$x_1 \geq 0.50$$
$$x_2 \geq 0.50$$
$$x_3 \geq 0.50$$
$$x_4 \geq 0.50$$

5. The budget is limited to $1500:

$$400x_1 + 200x_2 + 75x_3 + 100x_4 \leq 1500$$

The complete LP model for the example is presented as follows:

$$\text{Maximize}: \; x_1 + x_2 + x_3 + x_4 = 5$$

The optimal solution to the problem is shown in Table 4.2. Most of the conference time must be allocated to the seminar (2.20 days).

The expected annual cost savings due to this activity is $7,040,000. That is, 2.20 days × $3,200,000/year/day. Management games are the second most important activity. A total of 1.8 days for management games will yield annual cost savings of $3,600,000. Fifty percent of the remaining time (0.5 day) should be devoted to laboratory work, which will result in annual cost savings of $1,000,000. Case studies also require 0.5 day with a resulting annual savings of $200,000. The total annual savings, if the LP solution is implemented, is $11,840,000. Thus, an investment of $1,500 in management training for the personnel can generate annual savings of $11,840,000, a huge rate of return on investment!

RESOURCE COMBINATION FORMULATION

This example illustrates the use of the LP for energy resource allocation. Suppose an industrial establishment uses energy for heating, cooling, and lighting. The required amount of energy is presently being obtained from conventional electric power and natural gas. In recent years, there have been frequent shortages of gas, and there is a pressing need to reduce the consumption of conventional electric power. The director of the energy management department is considering a solar energy system as an alternate source of energy. The objective is to find an optimal mix of three different sources of energy to meet the plant's energy requirements. The three energy sources are:

- Natural gas
- Conventional electric power
- Solar power

It is required that the energy mix yield the lowest possible total annual cost of energy for the plant. Suppose a forecasting analysis indicates that the minimum kilowatt-hour (kwh) needed per year for heating, cooling, and lighting are 1,800,000, 1,200,000, and 900,000 kwh, respectively. The solar energy system is expected to supply at least 1,075,000 kwh annually. The annual use of conventional electric power must be at least 1,900,000 kwh due to a prevailing contractual agreement for energy supply. The annual consumption of the contracted supply of gas must be at least 950,000 kwh. The cubic foot unit for natural gas has been converted to kwh (1 ft³ of gas = 0.3024).

The respective rates of $6, $3, and $2 per kwh are applicable to the three sources of energy. The minimum individual annual savings desired are $600,000 from solar power, $800,000 from conventional electric power, and $375,000 from natural gas. The savings are associated with the operating and maintenance costs. The energy cost per kwh is $0.30 for conventional electric power, $0.20 for natural gas, and $0.40 for solar power. The initial cost of the solar energy system has been spread over its useful life of 10 years with appropriate cost adjustments to obtain the rate per kwh. The problem data is summarized in Table 4.3. If we let x_{ij} be the kwh used from source i for purpose j, then we would have the data organized as shown in Table 4.4.

The optimization problem involves the minimization of the total cost function, Z. The mathematical formulation of the problem is presented as follows:

$$\text{Minimize}: Z = 0.4 \sum_{j=1}^{3} x_{1j} + 0.3 \sum_{j=1}^{3} x_{2j} + 0.2 \sum_{j=1}^{3} x_{3j}$$

TABLE 4.3
Energy Resource Combination Data

Energy Source	Supply (1000s kwh)	Savings (1000s $)	Unit Savings ($/kwh)	Unit Cost ($/kwh)
Solar Power	1,075	600	6	0.40
Electric Power	1,900	800	3	0.30
Natural Gas	950	375	2	0.20

TABLE 4.4
Tabulation of Data for the LP Model

| Energy Source | Type of Use | | | Constraint |
	Heating	Cooling	Lighting	
Solar Power	X_{11}	X_{12}	X_{13}	$\geq 1{,}075K$
Electric Power	X_{21}	X_{22}	X_{23}	$\geq 1{,}900K$
Natural Gas	X_{31}	X_{32}	X_{33}	$\geq 950K$
Constraint	$\geq 1{,}800$	$\geq 1{,}200$	≥ 900	

TABLE 4.5
The LP Solution to the Resource Combination Example

| Energy Source | Type of Use | | |
	Heating	Cooling	Lighting
Solar Power	1,075	0	0
Electric Power	750	250	900
Natural Gas	0	950	0

$$\text{Subject to:} \quad x_{11} + x_{21} + x_{31} \geq 1{,}800$$

$$x_{12} + x_{22} + x_{32} \geq 1{,}200$$

$$x_{13} + x_{23} + x_{33} \geq 900$$

$$6\left(x_{11} + x_{12} + x_{13}\right) \geq 600$$

$$3\left(x_{21} + x_{22} + x_{23}\right) \geq 800$$

$$2\left(x_{31} + x_{32} + x_{33}\right) \geq 375$$

$$x_{11} + x_{12} + x_{13} \geq 1{,}075$$

$$x_{21} + x_{22} + x_{23} \geq 1{,}900$$

$$x_{31} + x_{32} + x_{33} \geq 950$$

$$x_{ij} \geq 0, \quad i, j = 1, 2, 3$$

The solution to this example is presented in Table 4.5. The table shows that solar power should not be used for cooling and lighting if the lowest cost is to be realized. The use of conventional electric power should be spread over the three categories of use. The solution indicates that natural gas should be used for cooling purposes. In pragmatic terms, this LP solution may have to be modified before being implemented on the basis of the prevailing operating scenarios and the technical aspects of the units involved.

RESOURCE REQUIREMENT ANALYSIS

Activity–resource assignment combinations provide opportunities for finding the best allocation of resources to meet project goals within the prevailing constraints in the project environment (Badiru, 1993, 2019). Suppose a manufacturing project requires that a certain number of workers be assigned to a workstation. The workers produce identical units of the same product. The objective is to determine the number of workers to be assigned to the workstation in order to minimize the total production cost per shift. Each shift is 8 hours long. Each worker can be assigned a variable number of hours and/or variable production rates to work during a shift. Four different production rates are possible: *slow rate, normal rate, fast rate,* and *high-pressure rate.* Each worker is capable of working at any of the production rates during a shift. The total number of work hours available per shift is determined by multiplying the number of workers assigned by the 8 hours available in a shift.

There are variable costs and percent defective associated with each production rate. The variable cost and the percent defective increase as the production rate increases. At least 450 units of the product must be produced during each shift. It is assumed that the workers' performance levels are identical. The production rates (r_i), the respective costs (c_i), and percent defective (d_i) are presented as follows:

Operating Rate 1 (Slow)
  $r_1 = 10$ units/h
  $c_1 = \$5/h$
  $d_1 = 5\%$
Operating Rate 2 (Normal)
  $r2 = 18$ units/h
  $c2 = \$10/h$
  $d2 = 5\%$
Operating Rate 3 (Fast)
  $r3 = 30$ units/h
  $c3 = \$15/h$
  $d3 = 12\%$
Operating Rate 4 (High Pressure)
  $r4 = 40$ units/h
  $c4 = \$25/h$
  $d4 = 15\%$

Let x_i represent the number of hours worked at production rate i.
Let n represent the number of workers assigned.
Let u_i represent the number of good units produced at operation rate i.
$u_1 = (10 \text{ units/h}) \cdot (1 - 0.05) = 9.50 \text{ units/h}$
$u_2 = (18 \text{ units/h}) \cdot (1 - 0.08) = 16.56 \text{ units/h}$
$u_3 = (30 \text{ units/h}) \cdot (1 - 0.12) = 24.40 \text{ units/h}$
$u_4 = (40 \text{ units/h}) \cdot (1 - 0.15) = 34.00 \text{ units/h}$

$$\text{Minimize}: z = 5x_1 + 10x_2 + 15x_3 + 25x_4$$

$$x_1 + x_2 + x_3 + x_4 \leq 8n$$
$$\text{Subject to}: \quad 9.50x_1 + 16.56x_2 + 25.40x_3 + 34.00x_4 \leq 450$$
$$x_1, x_2, x_3, x_4 \geq 0$$

The solution will be obtained by solving the LP model for different values of n. A plot of the minimum costs versus values of n can then be used to determine the optimum assignment policy. The complete solution can be explored further by interested readers.

INTEGER PROGRAMMING FORMULATION

Integer programming is a restricted model of the LP that permits only solutions with integer values of the decision variables. Suppose we are interested in minimizing the project completion time while observing resource limitations and job precedence relationships. The basic assumption is that once a job starts it has to be completed without interruption. We can construct several different integer programming models for the problem, all of which will give the same optimal solution but whose execution times will differ considerably. An efficient integer programming model for the problem should use as few integer variables as possible.

Define variables as:

$$x_{ij} : 1 \text{ if job } i \text{ starts in period } j; \quad 0, \text{ otherwise}$$
$$t_p : \text{completion time of the project}$$

Only x_{ij}s are restricted integers. For each job i, one can determine the early start and latest start times, ES_i and LS_i, respectively. Therefore, assuming that there are n jobs in the project, we have $1 \leq i \leq n$ for each i, then we have $ES_i \leq j \leq LS_i$. Let t_i denote the duration of job i. Resource availability constraints can be smartly handled by defining a vector V_{ij} which has 0s everywhere except positions $j, j + 1$, ..., $j + t_i - 1$, where it has 1s. It indicates the time period where job i uses the resource assuming that $x_{ij} = 1$. Let r_i and R_j be the resource required by job i and the resource available on day j, respectively. Let R be a row vector containing R_j.

Then, the integer programming model for the scheduling problem with limited resource can be defined as:

Minimize t_p

Subject to:

$$\sum_{j=ES_i}^{LS_i} x_{ij} = 1 \quad \forall i = 1, \ldots, n \tag{4.1}$$

$$-\sum_{j=ES_i}^{LS_i} jx_{ij} + \sum_{j=ES_k}^{LS_k} jx_{kj} \leq t_i \quad \forall k \in S(i) \tag{4.2}$$

$$\forall i = 1,\ldots,n$$

$$t_p - \sum_{j=ES_i}^{LS_i} jx_{ij} \geq t_i - 1 \quad \forall i \text{ with } S(i) = \phi \tag{4.3}$$

$$\sum_{i=1}^{n} \sum_{j=ES_i}^{LS_i} x_{ij} r_i V_{ij} \leq R$$

$$x_{ij} = 0,1 \quad \forall i = ES_i,\ldots,LS_i \tag{4.4}$$

$$\forall i = 1,\ldots,n$$

where $S(i)$ is the set of immediate successor jobs of job i.

Equation 4.1 indicates that each job must start on the same day. Equation 4.2 collectively makes sure that a job cannot start until all of the predecessor jobs are completed. Equation 4.3 determines the project completion time t_p. The project is completed after all the jobs without any successors are completed. The last set of equations makes sure that daily resource requirements are met. The indicator variable x_{ij} is restricted to the values 0 and 1.

The previous integer programming model can be solved using LINDO computer code and declaring x_{ij}s as binary variables. The code uses the branch and bound method of integer programming to solve the problem. Readers interested in the full details of the optimization modeling, examples, and solutions may refer to Badiru and Pulat (1995).

CONCLUSION

To get the best out of any system, we must think in terms of the best combination of goals, constraints, and resources. Only a mathematical representation and an optimization technique can handle the enormous combination of factors. Thus, the optimization techniques and formulations presented in this chapter offer a pathway to getting the most out of a supply chain, particularly in a case of disruptions, such as those caused by the COVID-19 pandemic.

REFERENCES

Badiru, A. B. (1993), "Activity resource assignments using critical resource diagramming," *Project Management Journal*, Vol. 14, No. (3), pp. 15–21.

Badiru, A. B. (2019), *Project Management: Systems, Principles, and Applications*, 2nd Edition, Taylor & Francis Group/CRC Press, Boca Raton, FL.

Badiru, A. B. and Simin Pulat, P. (1995), *Comprehensive Project Management: Integrating Optimization Models, Management Principles, and Computers*, Prentice Hall, Upper Saddle River, NJ, pp. 162–209.

5 Learning Curve in Production and Supply Chain Systems[1]

RESEARCH SYNOPSIS

This chapter presents an analytical research of learning curves in production and supply chain systems. The outputs of production systems are the elements that go into the supply chain. Learning curves are widely used in the design of production systems because the effect of learning helps to reduce the cost per unit as production runs increase. So, there is a good connectivity between learning curves, production outputs, and the supply chain. Whether the outputs are physical products, needed services, or desired results, the same concepts of learning curves are applicable to the supply chain.

In this reprinted journal article, the authors investigate production cost estimates to identify and model modifications for a prescribed learning curve. The research created a new learning curve for production processes that incorporates a new model parameter. The new parameter allows for a steeper learning curve at the beginning of production and a flattening effect near the end of production. The new model examines the learning rate as a decreasing function over time as opposed to a constant rate that is frequently used. The purpose of this research is to determine whether a new learning curve model could be implemented to reduce the error in cost estimates for production processes. A new model was created that mathematically allows for a "flattening effect," which typically occurs later in the production process. This model was then compared to Wright's learning curve, which is a popular method used by many organizations today. The results showed a statistically significant reduction in error through the measurement of the two error terms, Sum of Squared Errors and Mean Absolute Percentage Error.

INTRODUCTION

Many manufacturing firms today operate in a fiscally constrained and financially conscious environment. Managers throughout these organizations are expected to maximize the utility from every dollar as budgets and profit margins continue to shrink. Increased financial scrutiny adds greater emphasis on the accuracy of program and project management cost estimates to ensure acquisition programs are sufficiently funded. Cost estimating models and tools used by organizations must be evaluated for their relevance and accuracy to ensure reliable cost estimates.

Many of the cost estimating procedures for learning curves were developed in the 1930s (Wright, 1936) and are still in use today as a primary method to model learning. As automation and robotics increasingly replace human touch labor in the manufacturing process, the current 80-year-old learning curve model may no longer provide the most accurate approach for estimates. New learning curve methods that incorporate automated production and other factors that lead to reduced learning should be examined as an alternative for cost estimators in the acquisition process.

Since Wright's (1936) original learning curve model was developed, researchers have found other functions to model learning within the manufacturing process (Carr, 1946; Chalmers and DeCarteret, 1949; Crawford, 1944; DeJong, 1957; Towill, 1990; Towill and Cherrington, 1994). The purpose of this research is to address a gap in the literature that fails to account for the nonconstant rate of learning, which results in a flattening effect at the end of production cycles.

We investigate learning curve estimating methodology, develop learning curve theory, and pursue the development of a new estimation model that examines learning at a nonconstant rate.

This research identifies and models modifications to a learning curve model such that the estimated learning rate is modeled as a decreasing learning rate function over time, as opposed to a constant learning rate that is currently in use. Wright's (1936) learning curve model in use today mathematically states that for every doubling of units there will be a constant gain in efficiency. For example, if a manufacturer observes a 10% reduction in labor hours in the time to produce unit 10 from the time to produce unit 5, then it should expect to see the same 10% reduction in labor hours in the time to produce unit 20 from the time to produce unit 10. We propose that more accurate cost estimates would result if more flexible exponents were taken into consideration in developing the learning curve model. The proposed general modification would take the form:

$$Cost(x) = Ax^{f(x)} \tag{5.1}$$

where
$Cost(x)$ = cumulative average cost per unit
A = theoretical cost to produce the first unit
x = cumulative number of units produced
$f(x)$ = learning curve effect as a function of units produced

The exponent function in Equation 5.1 is explored in this article. Figure 5.1 demonstrates the phenomena this research examines.

To address this research gap, our study aims to model a function that has the added precision of diminishing learning effects over time by introducing a learning curve decay factor that more closely models actual production cycle learning. We accomplish this by developing a new learning curve model that minimizes the amount of error compared to current estimation models. Learning curves, specifically when estimating the expected cost per unit of complex manufactured items

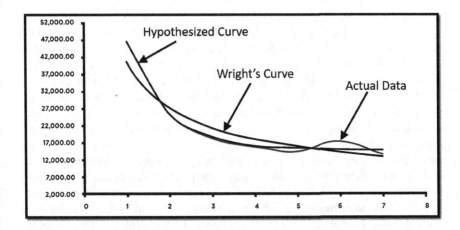

FIGURE 5.1   Learning Curve Depiction: Comparative Curves.

such as aircraft, are frequently modeled with a mathematical power function. The intent of these models is to capture the expected reduction in costs over time due to learning effects, particularly in areas with a high percentage of human touch labor. Typically, as production increases, manufacturers identify labor efficiencies and improve the process. If labor efficiencies are identified, it translates to unit cost savings over time. The general form of the learning curve model frequently used today is based on Wright's theory and is shown in Equation 5.2. Note that the structure of the exponent b ensures that as the number of units produced doubles, the cost will decrease by a given percentage defined as the learning curve slope (LCS). For example, when the LCS is 0.8, then the cost per unit will decrease by 80% between units 2 and 4.

$$Cost(x) = Ax^b \qquad (5.2)$$

where
   $Cost(x)$ = cumulative average cost per unit
   A = theoretical cost to produce the first unit
   x = cumulative number of units produced

$$b = \frac{\ln Learning\ Curve\ Slope}{\ln 2}$$

The cost of a particular production unit is modeled as a power function that decreases at a constant exponential rate. The problem is that the rate of decrease is not likely to be constant over time. We propose that the majority of cost improvements are to be found early in the production process, and fewer revelations are made later as the manufacturer becomes more familiar with the process. As time progresses, the production process should normalize to a steady state and additional cost reductions prove less likely. For relatively short production runs, the basic form of the learning curve may be sufficient because the hypothesized efficiencies will not have time to materialize. However, when estimating production runs over longer periods of time, the basic learning curve could underestimate the unit costs of those furthermost in the future (Godin and Warner, 2010: The Economist, 2012). The underestimation occurs because the model assumes a constant learning rate, while actual learning would diminish, causing the actuals to be higher than the estimate. Current models may underestimate a significant amount when dealing with high unit cost items such as those in major acquisition programs; a small error in an estimate can be large in terms of dollars. Through the use of curve fitting techniques, a comparison can be made to determine which model best predicts learning within a particular production process. The remainder of this article is organized as follows. A literature review of the most common learning curve processes is presented in the next section, followed by methodology and model formulation. We then provide an in-depth analysis of the learning curve models, followed by future research directions, conclusions, and limitations of this research.

LITERATURE REVIEW

Learning curve research dates back to 1936, when Theodore Paul Wright published the original learning curve equation that predicted the production effects of learning. Wright recognized the mathematical relationship that exists between the time it takes for a worker to complete a single task and the number of times the worker had previously performed that task (Wright, 1936). The mathematical relationship developed from this hypothesis is that as workers complete the same process, they get better at it. Specifically, Wright realized that the rate at which they get better at that task is constant. The relationship between these two variables is as follows: as the number of units produced doubles, the worker will do it faster by a constant rate. He proposed that this relationship takes the form of:

$$F = N^x \ or \ x = \frac{\log F}{\log N};$$

"where F = a factor of cost variation proportional to the quantity N. The reciprocal of F then represents a direct percent variation of cost vs. quantity" (Wright 1936). The relationship between these variables can be modified to predict the expected cost of a given unit number in production by multiplying the factor of cost variation by the theoretical cost of the first unit produced – this relationship was stated in Equation 5.2 and is shown in Figure 5.2. It is a log linear relationship through an algebraic manipulation. The logarithmic form of this equation (taking the natural log of both sides of the equation) allows practitioners to run linear regression analysis on the data to find what slope best fits the data using a straight line (Martin, 2019).

The goal of using learning curves is to increase the accuracy of cost estimates. Having accurate cost estimates allow an organization to efficiently budget while

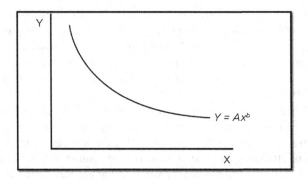

FIGURE 5.2 Wright's Learning Curve Model.

providing as much operational capability as possible because it can allocate resources to higher priorities. While the use of learning curves focuses on creating accurate cost estimates, learning curves often use the number of labor hours it takes to perform a task. When Wright originated the theory, he proposed the output in terms of time to produce, not production cost. However, many organizations perform learning curve analysis on both production cost and time to produce, depending on the data available. Nevertheless, labor hour cost is relevant because it is based on factors such as labor rates and other associated values. The use of labor hours in learning curve development allows a common comparison over time without the effects of inflation convoluting the results. However, the same goal can be achieved by using inflation-adjusted cost values.

Wright's model has been compared to some of the more contemporary models that have surfaced in recent years since the original learning curve theory was established (Moore et al., 2015). Moore compared the Stanford-B, DeJong, and the S-Curve models to Wright's model to see if any of these functions could provide a more accurate estimate of the learning phenomenon. Both the DeJong and the S-Curve models use an incompressibility factor in the calculation. Incompressibility is a factor used to account for the percentage of automation in the production process. Values of the incompressibility factor can range from 0 to 1 where 0 is all touch labor and 1 is complete automation. Moore found that when using an incompressibility factor between 0 and 0.1, the DeJong and S-Curve models were more accurate (Moore et al., 2015). In other words, when a production process had very little automation and high amounts of touch labor, the newer learning curve models tended to be more accurate. For all other values of incompressibility, Wright's model was more accurate.

More recently, Johnson (2016) proposed that a flattening effect is evident at the end of the production process where learning does not continue to occur at a constant rate near the end of a production cycle. Using the same models as Moore, Johnson explored the difference in accuracy between Wright's model and contemporary models early in the production process versus later in the production process. He had similar findings to Moore in that Wright's model was most accurate except in cases where the incompressibility factors were extremely low. When the incompressibility factor is low, more touch labor is involved in the process allowing for the possibility of additional learning to occur. He also found that Wright's learning curve was more accurate early in the production process, whereas the DeJong and S-Curve models were more accurate later in the production process (Johnson, 2016).

Another key concept in learning curve estimation and modeling is the idea of a forgetting curve (Honious et al., 2016). A forgetting curve explains how configuration changes in the production process can cause a break in learning, which leads to loss of efficiency that had previously been gained. When a configuration change occurs, the production process changes. Changes may include factors such as using different materials, different tooling, adding steps to a process, or might even be attributed to workforce turnover. The new process affects how workers complete their tasks and causes previously learned efficiencies to be lost. If manufacturers fail to take these breaks into account, they may underestimate the total effort needed to produce a product. Honious et al. (2016) found that configuration changes significantly changed the learning curve, and that the new LCS was steeper than the previous steady slope prior to a configuration change. The distinction between pre- and post-configuration change is important to ensure that both types of effects are taken into account.

The International Cost Estimating and Analysis Association (ICEAA) published learning curve training material in 2013. While presenting the basics of learning curve theory, it also presented some rules of thumb for learning. The first rule is that learning curves are steepest when the production process is touch labor intensive. Conversely, learning curves are the flattest when the production process is highly automated (ICEAA, 2013). Another key piece of information is that adding new work to the process can affect the overall cost. The ICEAA states that this essentially adds a new curve for the added work, which increases the original curve by the amount of the new curve (ICEAA, 2013). The equation is as follows:

$$Cost(x) = A_1 x^{b_1} + A_2 (x - L)^{b_2} \qquad (5.3)$$

where

$Cost(x)$ = cumulative average cost per unit
A_1 = theoretical cost to produce the first unit prior to addition of new work
x = cumulative number of units produced
L = last unit produced before addition of new work
A_2 = theoretical cost to produce the first unit after addition of new work
$b_1 = \dfrac{\ln Learning\ Curve\ Slope\ prior\ to\ additional\ work}{\ln 2}$
$b_2 = \dfrac{\ln Learning\ Curve\ Slope\ after\ additional\ work}{\ln 2}$ (typically same as b_1)

Equation 5.3 is important to consider when generating an estimate after a major configuration change or engineering change proposal (ECP). For example, while producing the eighth unit of an aircraft, the customer realizes they need to drastically change the radar on the aircraft. Learning has already taken place on the first eight aircraft; the new radar has not yet been installed, and therefore no learning has taken place. To accurately take into account the new learning, the radar would be treated as a second part to the equation, ensuring that we account for the learning on the eight aircraft while also accounting for no learning on the new radar.

Lastly, Anderlohr (1969) and Mislick and Nussbaum (2015) wrote about production breaks and the effects they have on a learning curve. These production breaks can cause a direct loss of learning, which can fully or partially reset the learning curve. For example, a 50% loss of learning would result in a loss of half of the cost reduction that has occurred (ICEAA, 2013). This information is important when analyzing past data to ensure that breaks in production are accounted for.

Thus far, we have laid out the fundamental building blocks for learning curve theory and how they might apply in a production environment. Wright's learning curve formula established the method by which many organizations account for learning during the procurement process. Following Wright's findings, other methods have emerged that account for breaks in production, natural loss of learning over time, incompressibility factors, and half-life analysis (Benkard, 2000). This article adds to the discussion by examining the flattening effect and how various models predict learning at different points in the production process.

When examining learning curve theory and the effects learning has on production, it is critical to understand the production process being estimated. Since Wright established learning curve theory in 1936, factory automation and technology have grown tremendously and continue to grow. Contemporary learning curve methods try to account for this automation. To get the best understanding, we examine the aircraft industry, specifically how it behaves in relation to the rest of the manufacturing industry.

The aircraft industry has relatively low automation (Kronemer and Henneberger, 1993), especially compared to other industries. Kronemer and Henneberger (1993) state that the aircraft industry is a fairly labor-intensive process with relatively little reliance on automated production techniques, despite it being a high-tech industry. Specifically, they list three main reasons why manufacturing aircraft is so labor intensive. First, aircraft manufacturers usually build multiple models of the same aircraft, typically for the commercial sector alone. These different aircraft models mean different tooling and configurations are needed to meet the demand of the customer. Second, aircraft manufacturers deal with a very low unit volume when compared to other industries in manufacturing. The final reason for low levels of automation is the fact that aircraft are highly complex and have very tight tolerances. To attain these specifications, manufacturers must continue to use highly skilled touch laborers or spend extremely large amounts of money on machinery to replace them (Kronemer and Henneberger, 1993). For these reasons, we should typically see or use low incompressibility factors in the learning curve models when estimating within the aircraft industry.

Although the aircraft industry remains largely unaffected by the shift to machine production from human touch labor, many industries are seeing a rise in the percentage of manufacturing processes that are automated. In a *Wall Street Journal* article posted in 2012, the author showed how companies have been increasing the amount of money spent on machines and software while spending less on manpower. They proposed that part of the reason behind this shift was a temporary tax break "that allowed companies in 2011 to write off 100% of investments in the first year" (Aeppel, 2012). Tax breaks combined with extremely low interest rates provided industry with incentive to invest in future production. Investment in production technology increases the incompressibility factor that should be used when estimating the effects of learning. In a separate article for the *Wall Street Journal*, Kathleen Madigan also pointed out the increase in spending on capital investments in relation to labor. She stated that "businesses had increased their real spending on equipment and software by a strong 26%, while they have added almost nothing to their payrolls" (Madigan, 2011).

MODEL FORMULATION

Before we can begin the process of developing a new learning curve equation, we need to examine the characteristics of the curve we expected to best fit the data. Our hypothesis is that a learning curve whose slope decreases over time would fit the data better than Wright's curve. To adjust the rate at which the curve flattens, the b value from Wright's learning curve, or the exponent in the power function, needs to be adjusted. Specifically, to make the curve move in a flatter direction, the exponent in the power curve must decrease as the number of units produced increases. Initially, we modified Wright's existing formula by dividing the exponent by the unit number as shown in Equation 5.4.

$$Cost(x) = Ax^{b/x} \qquad (5.4)$$

where
$Cost(x)$ = cumulative average cost per unit
A = theoretical cost of the first unit
x = cumulative number of units produced
b = Wright's learning curve constant as described in Equation 5.2

Using Wright's learning curve, b is a negative constant that has a larger magnitude for larger amounts of learning (i.e., as the LCS decreases, b becomes more negative). Therefore, in Equation 5.4, when b is divided by x, the amount of learning is reduced. In fact, the flattening effect is fairly drastic. For example, when applying Equation 5.4, a standard 80% Wright's learning curve exhibits 90% learning by the second unit and flattens to 97% by the fourth unit. To implement a learning curve that has the flexibility to not flatten as quickly, we instead divide b

by 1 + x/c where c is a positive constant (see Equation 5.5). The term 1 + x/c is always greater than 1 and increases as x increases; therefore, a flattening effect always occurs (i.e., learning decreases as the number of units produced increases). The choice of the constant c is critical in determining how quickly the learning decreases. For example, when $c = 4$, a standard 80% Wright's learning curve exhibits 86% learning by the second unit and approximately 89% learning by the fourth unit. For the same standard 80% curve when $c = 40$, the learning decreases to 80.9% by the second unit and to 81.6% by the fourth unit. The new equation (which we also refer to as Boone's learning curve hereafter) took the form:

$$Cost(x) = Ax^{b/\left(1+\frac{x}{c}\right)} \tag{5.5}$$

where
    $Cost(x)$ = cumulative average cost per unit
    A = theoretical cost of the first unit
    x = cumulative number of units produced
    b = Wright's learning curve constant as described in Equation 5.2
    c = decay value (positive constant)

The function that modifies the traditional learning curve exponent in Equation 5.5 – i.e., 1 + x/c – has a key characteristic – that ensures that the rate of learning associated with traditional learning curve theory decreases as each additional unit is produced. Specifically, 1 + x/c is always greater than 1 since x/c is always positive. Note that c is an estimated parameter and x increases as more units are produced, so the term x/c is decreasing. When c is large, Boone's learning curve would effectively behave like Wright's learning curve. For example, if the fitted value of c is 5,000, then 1 + x/c equals 1.0002 after the first unit has been produced and 1.004 after the 20th unit has been produced. This equates to a decrease in the learning rate of the traditional theory (i.e., b) to less than 0.07%. More formally, as c goes to infinity, $b/(1 + x/c)$ goes to b.

Note that the previous discussion assumed that b was the same value for both Wright's and Boone's learning curve to help demonstrate the flattening effect. In practice, nothing precludes each of the learning curves from having different b values. For instance, if we desire a learning curve that possesses more learning early in production and less learning later in production (compared to Wright's curve), then the b parameters could be different. This is shown in Figure 5.1. In this case, Boone's curve would have a b value less than Wright's curve (i.e., a more negative value representing more learning). Then the flattening effect of dividing by 1 + x/c as production increases would eventually result in a curve with less learning than Wright's curve. For example, consider an 80% Wright's learning curve and a Boone's learning curve that initially has 70% learning and a decay value of 8; by the eighth production unit, Boone's curve would be at 82% learning.

POPULATION AND SAMPLE

To test the new learning curve in Equation 5.5, we looked at quantitative data from several DoD airframes to gain a comprehensive understanding of how learning affects the cost of lot production. The costs used in this analysis are the direct lot costs and exclude costs for items such as Research, Development, Test, and Evaluation (RDT&E), support items, and spares. These data specifically include Prime Mission Equipment (PME) only as these costs are directly related to labor and can be influenced directly through learning. To ensure we are comparing properly across time, we used inflation and rate-adjusted PME cost data for each production lot of the selected aircraft systems. The PME cost data were adjusted using escalation rates for materials using the Office of Secretary of Defense (OSD) rate tables, when applicable. We used data from fighter, bomber, and cargo aircraft, as well as missiles and munitions. This diverse dataset allowed comparison among multiple systems in different production environments.

DATA COLLECTION

Data used were gathered from the Cost Assessment Data Enterprise (CADE). The CADE is a resource available to the DoD cost analysts that stores historical data on weapon systems. Some of the older data also came from a DoD research library in the form of cost summary reports. The data used can be broken out by Work Breakdown Structure (WBS) or Contract Line Item Number (CLIN). For this research, the PME cost data were broken out by WBS element, then rolled up into top line, finished product elements, and used for the regression analysis. In total, 46 weapon system platforms were analyzed (see Table provided in Boone et al., 2021).

ANALYSIS

Regression analysis was used to test which learning curve model was most accurate in estimating the data. The goal is to minimize the sum of squared errors (SSE) in the regression to examine how well a model estimates a given set of data. The SSE is calculated by taking the vertical distance between the actual data point (in this case lot midpoint PME cost) and the prediction line (or estimate) (Mislick and Nussbaum, 2015). This error term is then squared and the sum of these squared error terms is the value for comparing which model is a more accurate predictor. However, since an extra parameter is available in fitting the regression for the new model, it should be able to maintain or decrease the SSE in most cases. As previously mentioned, as the decay parameter in Equation 5.5 approaches infinity, Boone's learning curve approaches Wright's learning curve formula. With this in mind, we also examined the Mean Absolute Percentage Error (MAPE). The MAPE takes the same error term as the SSE calculation but then divides it by the actual value; then the mean of the absolute value of these modified error terms is calculated. By examining the error in terms of a percentage, comparisons between different types and sizes of systems are more robust. If Boone's curve reduces both the SSE and MAPE when compared to the SSE and MAPE of Wright's curve, it would indicate that the new model may be better suited for modeling learning and the associated costs.

As stated previously, Wright's learning curve is suitable for a log-log model. A log-log model is used when a logarithmic transformation of both sides of an equation results in a model that is linear in the parameters. As Wright proposed, this linear transformation occurs because learning happens at a constant rate throughout the production cycle. If learning happens at a nonconstant rate (as in Boone's learning curve), then the curve in log-log space would no longer be linear. This constraint means typical linear regression methods would not be suitable for estimating Boone's learning curve; therefore, we had to use nonlinear methods to fit these curves.

Specifically, we used the Generalized Reduced Gradient (GRG) nonlinear solver package in Excel to minimize the SSE by fitting the A, b, and c parameters from Equation 5.5. To use this solver, bounds for the three parameters had to be established. These are values that are easy to obtain for any dataset, as they are provided by Microsoft Excel when fitting a power function or by using the "linest()" function in Excel. We used this as a starting point because Wright's curve is currently used throughout the DoD. For the A variable, the lower bound was one-half of Wright's A and the upper bound was two times Wright's A. These values were used to give the solver model a wide enough range to avoid limiting the value but small enough to ease the search for the optimal values. Neither of these limits was found to be binding. For the exponent parameter b, we chose values between 3 and -3 times Wright's exponent value. In theory, the value of the exponent should never go above 0 due to positive learning leading to a decrease in cost, but in practice some datasets go up over time and we wanted to be able to account for those scenarios, if necessary. Again, these values between 3 and -3 times Wright's exponent value were never found to be binding limits for the model. Finally, for the decay parameter c, fitted values were bounded between 0 and 5,000; the 5,000 upper bound was found to be a binding constraint in the solver on several occasions. In practice, analysts could bound the value as high as possible to reduce error, but in the case of this research, we used 5,000 as no significant change was evidenced from 5,000 to infinity – relaxing this bound would have only further reduced the SSE for Boone's learning curve.

STATISTICAL SIGNIFICANCE TESTING

Once the SSE and MAPE values were calculated for each learning curve equation, we tested for significance to determine whether the difference between the error values for the two equations were statistically different. Specifically, we conducted t-tests on the differences in error terms between Wright's and Boone's learning curve equations. This t-test was conducted for both the SSE and MAPE values separately. A nonsignificant t-test indicates no statistically significant difference between the two learning curves.

ANALYSIS AND RESULTS

The Table (see Boone et al., 2021, p. 86) shows the SSE and MAPE values for both Wright's and Boone's learning curve for each system in the dataset. The last two columns are the percentage difference in the SSE and MAPE between the two learning curve methods. This percentage was calculated by taking the difference of Boone's error term minus Wright's error term divided by Wright's error term. Negative values represent programs where Boone's learning curve had less error than Wright's learning curve, and positive values represent programs where Wright's curve had less error than Boone's curve.

Based on this analysis, we observed that Boone's learning curve reduced the SSE in approximately 84% of programs and reduced the MAPE in 67% of programs. The mean reduction of the SSE and MAPE was 27% and 17%, respectively. As previously mentioned, these values were based on using both learning curve equations to minimize the SSE for each system in the dataset. This is a standard practice in the DoD as prescribed by the US Government Accountability Office (GAO, 2009) *Cost Estimating and Assessment Guide* when predicting the cost of subsequent units or subsequent lots.

We conducted additional tests to determine if a statistical difference existed between the means of both curve estimation techniques. On average, programs estimated using Boone's learning curve had a lower error rate (M = 4.73, SD = 2.15) than those estimated using Wright's learning curve (M = 8.64, SD = 4.55). Additionally, the difference between these two error rates expressed as a percentage and compared to a hypothesized value of 0 (no difference) was significant, t (46) = −4.87, p < .0001 and represented an effect of d = 1.10. We then applied the same test to the difference in the MAPE values from Boone's learning curve and Wright's learning curve. On average, programs estimated using Boone's learning curve had a lower MAPE value (M = .08, SD = .07) than those estimated using Wright's MAPE value (M = .10, SD = .06). The difference between these two estimates has a mean of −0.17, which translates to Boone's curve reducing the MAPE by 17% more on average. Additionally, the difference between these two error rates expressed as a percentage and compared to a hypothesized value of 0 (no difference) was significant, t(46) = −3.48, p < .0005 and represented an effect of d = .22. The results indicate that in both the SSE and MAPE, Boone's learning curve reduced the error, and that each of those values was statistically significant when using an alpha value of 0.05.

DISCUSSION

As stated previously, an average of a 27% reduction in the SSE resulted from among the 46 programs analyzed. These results were statistically significant. Also, a 17% reduction in the MAPE resulted from among the programs analyzed, which was also found to be statistically significant. Based on these results, we can conclude that Boone's learning curve equation was able to reduce the overall error in cost estimates for our sample. This information is critical to allow the DoD to calculate more accurate cost estimates and better allocate its resources. These conclusions help answer our three guiding research questions. Specifically, we were looking for the point where Wright's model became less accurate than other models. We found that adding a decay factor caused the learning curve to flatten out over time, which resulted in less error than Wright's model. Additionally, we found that Boone's learning curve was more accurate throughout the entire production process, not just during the tail end when production was winding down. Boone's learning curve was steeper during the early stages of production when it's hypothesized that the most learning occurs. Toward the end of the production process, Boone's curve flattens out more than Wright's curve, supporting our contention that learning toward the end of the production cycle yields diminishing returns. While Wright's curve assumes constant learning throughout the entire process, Boone's treats learning in a nonlinear fashion that slows down over time. By reducing the error in the estimates and properly allocating resources, the DoD could potentially minimize risk for all parties involved. The benefit of Boone's learning curve is accuracy in the estimation process. If labor estimates aren't accurate in the production process, risks escalate, such as schedule slip, cost overruns, and increased costs for all involved. Accuracy in the cost estimate should be the goal of both the contractor and government, thereby facilitating the acquisition process with better data.

LIMITATIONS

One limitation of this study is that all 46 of the weapon systems analyzed were US Air Force systems. While the list included many platforms spanning decades, we hesitate to draw conclusions outside of the US Air Force without further research and analysis. That said, we see no reason that our model wouldn't apply equally well in any aircraft production environment, both within and outside the DoD. Another limitation in this research is the use of PME cost as opposed to labor hours. Labor-hour data are not readily available across many platforms, which led to the use of PME cost. Contractor data provided to the government normally come in the form of lots, which is the lowest level tracked by cost estimators. To compare learning curves across multiple platforms, the same level of analysis is required to ensure a fair comparison. Future research should attempt to examine data at the individual level of analysis between systems and exclude those where only lot data are available. Because there are inherently less lots than units, this may affect how the equation behaves when applied at the unit level. For this research, we used the lot midpoint formula/method (Mislick and Nussbaum, 2015), but further research should be conducted to evaluate the performance of Boone's learning curve with unitary data. Finally, we only performed a comparison to Wright's learning curve since that is a primary method of estimation in the DoD. A comparison with other learning curve models may yield different results, although previous research found those curves were not statistically better than Wright's.

RECOMMENDATIONS FOR FUTURE RESEARCH

Data outside of the US Air Force should be examined to test whether this equation applies broadly to programs and not just to Air Force programs. Also, conducting the analysis with unitary data could confirm that this works for predicting subsequent units as well as subsequent lots, while reducing error over Wright's method. We also made an attempt to select weapon systems that had minimal automation in the production process. However, DeJong's Learning Formula is another derivation from Wright's original in which an incompressibility factor is introduced. The incompressibility factor represents the amount of automation in the production process, which limits how much learning can occur (Badiru et al., 2013). Other models such as the S-Curve model (Carr, 1946) and a more recent version (Towill, 1990; Towill and Cherrington, 1994) also account for some form of incompressibility. Additional research could also include modifications to Boone's formula to try and further reduce the error types listed in this research. Furthermore, fitting Boone's curve in this analysis was based on past data whereas cost estimates are used to project future costs. Therefore, future research should identify decay values for different types of weapon systems – similar to the way learning curve rates are established for different categories of programs. Lastly, further research could examine whether the incorporation of multiple learning curve equations at different points in the production process would be beneficial to reducing additional error in the estimates.

We developed a new learning curve equation utilizing the concept of learning decay. This equation was tested against Wright's learning equation to see which equation provided the least amount of error when looking at both the SSE and MAPE. We found that Boone's learning curve reduced error in both cases and that this reduction in error was statistically significant. Follow-on research in this field could lead to further discoveries and allow for broader use of this equation in the cost community.

LEARNING CURVE IMPLICATION FOR SUPPLY CHAINS

Although the methodology of this research centers on general production systems using Air Force data, it is directly applicable to the case of general relationships between a production system and a supply chain networks. The half-life learning curve model presented by Badiru (2012) is particularly relevant for dynamic technology-sensitive supply chains, where learning and relearning can change over a short period of time. An appreciation and incorporation of learning curves can help mitigate supply chain risks.

NOTES

1 Reprinted with permission from: "A Learning Curve Model Accounting for the Flattening Effect in Production Cycles" by Boone, E., Elshaw, J. J., Koschnick, C. M., Ritschel, J. D., and Badiru, A. B. (2021). *Defense Acquisition Research Journal*, Vol. 28, No. 1, Jan 2021, pp. 72–97, Issue 95 by DAU Press. DOI: https://doi.org/10.22594/10.22594/dau.20-850.28.01

REFERENCES

Aeppel, T. (2012, January 17), "Man vs. machine: behind the jobless recovery," *Wall Street Journal*: http://www.wsj.com/articles/SB10001424052970204468004577164710231081398

Anderlohr, G. (1969), "What production breaks cost," *Industrial Engineering*, Vol. 1, No. (9), pp. 34–36.

Badiru, A. B. (2012), "Half-life learning curves in the defense acquisition life cycle," *Defense Acquisition Research Journal*, Vol. 19, No. (3), pp. 283–308: https://www.afit.edu/BIOS/publications/HalflifeLearningCurvesinDefenseAcquisitionLife-CycleBadiruDARJ2012.pdf

Badiru, A., Elshaw, J., and Mack, E. (2013), "Half-life learning curve computations for airframe life-cycle costing of composite manufacturing," *Journal of Aviation and Aerospace Perspective*, Vol. 3, No. (2), pp. 6–37.

Boone, E., Elshaw, J. J., Koschnick, C. M., Ritschel, J. D., Badiru, Adedeji (2021). "A New Learning Curve Model Accounting for the Flattening Effect at the End of Production Cycles," *Defense Acquisition Research Journal*, Vol. 28, No. 1, Jan 2021, pp. 72–97.

Benkard, C. L. (2000). "Learning and forgetting: the dynamics of aircraft production," *American Economic Review*, Vol. 90, No. (4), pp. 1034–1054: https://www.aeaweb.org/articles?id=10.1257/aer.90.4.1034

Carr, G. W. (1946), "Peacetime cost estimating requires new learning curves," *Aviation*, Vol. 45, No. (4), pp. 220–228.

Chalmers, G., and DeCarteret, N. (1949), *Relationship for Determining the Optimum Expansibility of the Elements of a Peacetime Aircraft Procurement Program*, USAF Air Materiel Command, Wright-Patterson AFB, OH.

Crawford, J. R. (1944), *Estimating, budgeting and scheduling*, Lockheed Aircraft Co., Burbank, CA.

DeJong, J. R. (1957), "The effects of increasing skill on cycle time and its consequences for timestandards,"*Ergonomics*,Vol.1,No.(1),pp.51–60.doi:10.1080/00140135708964571

Godin, S. and Warner, A. (2010, January 25), How to produce like a linchpin by understanding your lizard brain (No. 214) [Audio podcast episode]. In *Recession-Proof Startups. Mixergy*: https://radiopublic.com/recessionproof-startups-mixergy-WolJm6/s1!087be

Honious, C., Johnson, B., Elshaw, J., and Badiru, A. (2016, May 4–5). *The impact of learning curve model selection and criteria for cost estimation accuracy in the DoD*. In Maj Gen C. Blake, USAF (Chair), *Proceedings of the 13th Annual Acquisition Research Symposium*, Monterey, CA: https://apps.dtic.mil/dtic/tr/fulltext/u2/1016823.pdf

International Cost Estimating and Analysis Association (2013), *Cost estimating body of knowledge* [PowerPoint slides], Torino, Italy: ICEAA: https://www.iceaaonline.com/cebok/, September 9–13, 2013

Johnson, B. J. (2016). *A comparative study of learning curve models and factors in defense cost estimating based on program integration, assembly, and checkout* [Master's thesis]. Air Force Institute of Technology: https://apps.dtic.mil/dtic/tr/fulltext/u2/1056447.pdf

Kronemer, A., and Henneberger, J. E. (1993), "Productivity in aircraft manufacturing," *Monthly Labor Review*, Vol. 16, No. (6), pp. 24–33. http://www.bls.gov/mfp/mprkh93.pdf

Madigan, K. (2011, September 28), "It's man vs. machine and man is losing," *Wall Street Journal*: https://blogs.wsj.com/economics/2011/09/28/its-man-vs-machine-and-man-is-losing/

The Economist (2012), "Making the future: How robots and people team up to manufacture things in new ways". *The Economist*: https://www.economist.com/special-report/2012/04/21/making-the-future, accessed 29 September 2021.

Martin, J. R. (2019), *What Is a Learning Curve?* Management and Accounting Web. Retrieved June 15, 2019 from http://maaw.info/LearningCurveSummary.htm

Mislick, G. K., and Nussbaum, D. A. (2015), *Cost Estimation: Methods and Tools*, Wiley: https://www.academia.edu/24430098/Cost_Estimation_Methods_and_Tools_by_Gregory_K_Mislick_and_Daniel_A_Nussbaum_1st_Edition

Moore, J. R., Elshaw, J. Badiru, A. B., and Ritschel, J. D. (2015, October), "Acquisition challenge: the importance of incompressibility in comparing learning curve models," *Defense Acquisition Research Journal*, Vol. 22, No. (4), pp. 416–449: https://www.dau.edu/library/arj/ARJ/ARJ75/ARJ75-ONLINE-FULL.pdf

Towill, D. R. (1990), "Forecasting learning curves," *International Journal of Forecasting*, Vol. 6, No. (1), pp. 25–38: https://www.sciencedirect.com/science/article/abs/pii/016920709090095S

Towill, D. R., and Cherrington, J. E. (1994), "Learning curve models for predicting the performance of AMT," *The International Journal of Advanced Manufacturing Technology*, Vol. 9, pp. 195–203. doi:10.1007/BF01754598

U.S. Government Accountability Office. (2009), *GAO Cost Estimating and Assessment Guide (GAO-09-3SP)*, Government Printing Office. https://www.gao.gov/new.items/d093sp.pdf

Wright, T. P. (1936), "Factors affecting the cost of airplanes," *Journal of the Aeronautical Sciences*, Vol. 3, No. (4), pp. 122–128: https://arc.aiaa.org/doi/10.2514/8.155

6 Supplier Selection Using Multiple Criteria Optimization

PREAMBLE

The performance of a supply chain is ultimately a function of the inputs going into the supply chain. These inputs originate from the supplier sources that produce physical products, desired services, or needed services. If the source can be optimized, the process can be improved, and the supply output can be enhanced. When supply disruptions occur, as we saw in the case of COVID-19 pandemic, the origin of problems can be from any point in the overall spectrum of the supply chain. It is in this regard that this chapter is dedicated to providing the quantitative methodology of optimizing supplier selection. This chapter is based on the approach found in Ravindran and Wadhwa (2009). Full details of the approach can be found in Wadhwa and Ravindran (2007), Ravindran and Wadhwa (2009), Ravindran (2008), Ravindran (2018), Ravindran and Warsing (2017), and all the references therein.

SUPPLIER SELECTION PROBLEM

A supply chain consists of a connected set of activities concerned with planning, coordinating, and controlling materials, parts, and finished good from supplier to customer (Ravindran and Wadhwa, 2009). The contribution of the purchasing function to the profitability of the supply chain has assumed greater proportions in recent years; one of the most critical functions of purchasing is selection of suppliers. For most manufacturing firms, the purchasing of raw material and component parts from suppliers constitutes a major expense. Raw material cost accounts for 40–60% of production costs for most US manufacturers. In fact, for the automotive industry, the cost of components and parts from outside suppliers may exceed 50% of sales (Ravindran and Wadhwa, 2009). For technology firms, purchased materials and services account for 80% of the total production cost. It is vital to the competitiveness of most firms to be able to keep the purchasing cost to a minimum. In today's competitive operating environment, it is impossible to successfully produce low-cost, high-quality products without good suppliers. A study carried out by the Aberdeen group (Ravindran and Wadhwa, 2009) found that more than 83% of the organizations engaged in outsourcing achieved significant reduction in purchasing cost, more than 73% achieved reduction in transaction cost, and over 60% were able to shrink sourcing cycles.

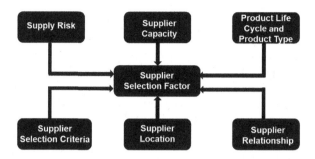

FIGURE 6.1   Supplier Selection Factors.

(Adapted from Ravindran and Wadhwa, 2009).

Supplier selection process is difficult because the criteria for selecting suppliers could be conflicting. Figure 6.1 illustrates the various factors which could impact the supplier selection process (Ravindran and Wadhwa, 2009). Supplier selection is a multiple criteria optimization problem that requires trade-off among different qualitative and quantitative factors to find the best set of suppliers. For example, the supplier with the lowest unit price may also have the lowest quality. The problem is also complicated by the fact that several conflicting criteria must be considered in the decision-making process.

Most of the times buyers have to choose among a set of suppliers by using some predetermined criteria such as, quality, reliability, technical capability, lead-times, etc., even before building long-term relationships. To accomplish these goals, two basic and interrelated decisions must be made by a firm. The firm must decide which suppliers to do business with and how much to order from each supplier. Weber et al. (Ravindran and Wadhwa, 2009) refer to this pair of decisions as the supplier selection problem.

Figure 6.2 illustrates the steps in the supplier selection process. The first step is to determine whether to *make or buy* the item. Most organizations buy those parts which are not core to the business or not cost-effective if produced in-house. The next step is to define the various criteria for selecting the suppliers. The criteria for

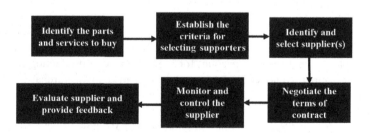

FIGURE 6.2   Supplier Selection Steps.

(Adapted from Ravindran and Wadhwa, 2009).

selecting a supplier of critical product may not be the same as a supplier of MRO items. Once a decision to buy the item is taken, the most critical step is selecting the right supplier. Once the suppliers are chosen, the organization has to negotiate terms of contract and monitor their performance. Finally, the suppliers have to be constantly evaluated, and the feedback should be provided to the suppliers.

SUPPLIER SELECTION MODELS

As mentioned earlier, the supplier selection activity plays a key role in cost reduction and is one of the most important functions of the purchasing department. Different mathematical, statistical, and game theoretical models have been proposed to solve the problem. References in Ravindran and Wadhwa (2009) provide an overview of supplier selection methods.

De Boer et al. (Ravindran and Wadhwa, 2009) stated that supplier selection process is made up of several decision-making steps as shown in Figure 6.3.

Problem Formulation

Decision support methods for problem formulation are those methods that support the purchasing manager in identifying the need for supplier selection; in this case, problem formulation involves determining what the ultimate problem is and why supplier selection is the better choice. According to a survey by Aissaoui et al (Ravindran and Wadhwa, 2009) and De Boer et al. (Ravindran and Wadhwa, 2009), there are no papers that deal with the problem formulation.

Criteria for Selection

Depending on the buying situation, different sets of criteria may have to be employed. Criteria for supplier selection have been studied extensively since the

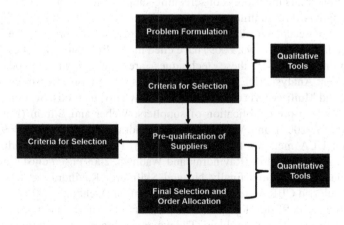

FIGURE 6.3 Supplier Selection Process.

(Adapted from Ravindran and Wadhwa, 2009).

1960s. Dickson's study (Ravindran and Wadhwa, 2009) was the earliest to review the supplier selection criteria. He identified 23 selection criteria with varying degrees of importance for the supplier selection process. Dickson's study was based on a survey of purchasing agents and managers. A follow-up study was done by Weber et al. (Ravindran and Wadhwa, 2009). They reviewed 74 articles published in the 1970s and 1980s and categorized them based on Dickson's 23 selection criteria. In their studies, net price, quality, delivery, and service were identified as the key selection criteria. They also found that under Just-In-Time manufacturing, quality and delivery became more important than net price. Supplier selection criteria may also change over time. Wilson (Ravindran and Wadhwa, 2009) examined studies conducted in 1974, 1982, and 1993 on the relative importance of the selection criteria. She found that quality and service criteria began to dominate price and delivery. She concluded that globalization of the market place and the increasingly competitive atmosphere had contributed to the shift.

De Boer et al. (Ravindran and Wadhwa, 2009) pointed out that there are only two known techniques for formulating supplier selection criteria. Mandal and Deshmukh (Ravindran and Wadhwa, 2009) proposed interpretive structural modeling (ISM) as technique based on group judgment to identify and summarize relationships between supplier choice criteria through a graphical model. Vokurka et al. (Ravindran and Wadhwa, 2009) developed an expert system for the formulation of the supplier selection criteria. Nominal Group Technique involving all the stakeholders of the supplier selection decision can also be used to identify the important supplier selection criteria (Ravindran and Wadhwa, 2009).

PRE-QUALIFICATION OF SUPPLIERS

Pre-qualification is the process of screening suppliers to identify a suitable subset of suppliers. Pre-qualification reduces the large set of initial suppliers to a smaller set of acceptable suppliers and is more of a sorting process. De Boer et al. (Ravindran and Wadhwa, 2009) have cited many different techniques for pre-qualification. Some of these techniques are Categorical Methods, Data Envelopment Analysis (DEA), Cluster Analysis, Case-based Reasoning (CBR) systems, and Multi-criteria decision-making (MCDM) methods. Several authors have worked on pre-qualification of suppliers. Weber and Ellram (Ravindran and Wadhwa, 2009) and Weber et al. (Ravindran and Wadhwa, 2009) have developed DEA methods for pre-qualification. Hinkel et al. (Ravindran and Wadhwa, 2009) and Holt (Ravindran and Wadhwa, 2009) used cluster analysis for pre-qualification, and finally Ng and Skitmore (Ravindran and Wadhwa, 2009) developed CBR systems for pre-qualification. Mendoza et al. (Ravindran and Wadhwa, 2009) developed a three phase multi-criteria method to solve a general supplier selection problem. The paper combines the Analytic Hierarchy Process (AHP) with goal programming (GP) for both pre-qualification and final order allocation.

FINAL SELECTION

Most of the publications in the area of supplier selection have focused on final selection. In the final selection step, the buyer identifies the suppliers to do business with and allocates order quantities among the chosen supplier(s). In reviewing the literature, there are basically two kinds of supplier selection problem, as stated by Ghodsypour and O'brien (Ravindran and Wadhwa, 2009):

- Single sourcing, which implies that any one of the suppliers can satisfy the buyer's requirements of demand, quality, delivery, etc.
- Multiple sourcing, which implies that there are some limitations in suppliers' capacity, quality, etc., and multiple suppliers have to be used.

Specifically, no one supplier can satisfy the buyer's total requirements and the buyer needs to purchase some part of its demand from one supplier and the other part from another supplier to compensate for the shortage of capacity or low quality of the first supplier. Several methods have been proposed in the literature for single sourcing as well as for multiple sourcing.

SINGLE SOURCING MODELS

Single sourcing is a possibility when a relatively small number of parts are procured (Ravindran and Wadhwa, 2009). Some of the methods used in single sourcing are:

- Linear Weighted Point: The Linear Weighted Point method is the most widely used approach for single sourcing. This approach uses a simple scoring method, which heavily depends on human judgment. Some of the references that discuss this approach include Wind and Robinson (Ravindran and Wadhwa, 2009), Zenz (Ravindran and Wadhwa, 2009), etc.
- Cost Ratio: Cost Ratio is a more complicated method in which the cost of each criterion is calculated as a percentage of the total purchased value, and a net adjusted cost for each supplier is determined. This approach is complex and needs a lot of financial information. This approach was proposed by Timmerman (Ravindran and Wadhwa, 2009).
- Analytic Hierarchy Process (AHP): AHP, developed by Saaty (Ravindran and Wadhwa, 2009) in the early 1980s, is a good tool to solve multiple criterion problems with finite alternatives. It is a powerful and flexible decision-making process to help the decision-maker set priorities and make the best decision when both qualitative and quantitative aspects of decision are to be considered. The AHP has been extensively used for supplier selection problem (Ravindran and Wadhwa, 2009).
- Total Cost of ownership (TCO): TCO is a method which looks beyond the price of a product to include other related costs like quality costs, technology costs, etc. Finally, the business is awarded to the supplier with lowest unit total cost. General Electric Wiring Devices have developed a total cost

supplier selection method that takes into account risk factors, business desirable factors, and measurable cost factors (Ravindran and Wadhwa, 2009). The TCO approach has also been used by Ellram (Ravindran and Wadhwa, 2009) and Degreave (Ravindran and Wadhwa, 2009).

MULTIPLE SOURCING MODELS

Multiple sourcing can offset the risk of supply disruption. In multiple sourcing, a buyer purchases the same item from more than one supplier. Mathematical programming is the most appropriate method for multiple sourcing decisions. It allows the buyers to consider different constraints while choosing the suppliers and their order allocation. Two types of mathematical programming models are found in the literature, single objective and multi-objective models.

Single Objective Models

Moore and Fearon (Ravindran and Wadhwa, 2009) stated that price, quality, and delivery are important criteria for supplier selection. They discussed the use of linear programming in the decision-making.

Gaballa (Ravindran and Wadhwa, 2009) applied mathematical programming to supplier selection in a real case. He used a mixed integer programming to formulate a decision-making model for the Australian Post Office. The objective for this approach is to minimize the total discounted price of allocated items to the suppliers. Anthony and Buffa (Ravindran and Wadhwa, 2009) developed a single objective linear programming model to support strategic purchasing scheduling (SPS). The linear model minimized the total cost by considering limitations of purchasing budget, supplier capacities, and buyer's demand. Price and storage cost were included in the objective function. The costs of ordering, transportation, and inspection were not included in the model. Bender et al. (Ravindran and Wadhwa, 2009) applied single objective programming to develop a commercial computerized model for supplier selection at IBM. They used mixed integer programming, to minimize the sum of purchasing, transportation, and inventory costs. Narasimhan and Stoynoff (Ravindran and Wadhwa, 2009) applied a single objective, mixed integer programming model to a large manufacturing firm in the Midwest, to optimize the allocation procurement for a group of suppliers. Turner (Ravindran and Wadhwa, 2009) presented a single objective linear programming model for British Coal. This model minimized the total discounted price by considering the supplier capacity, maximum and minimum order quantities, demand, and regional allocated bounds as constraints. Pan (Ravindran and Wadhwa, 2009) proposed multiple sourcing for improving the reliability of supply for critical materials, in which more than one supplier is used and the demand is split between them. The author used a single objective linear programming model to choose the best suppliers, in which three criteria are considered price, quality, and service. Seshadri et al. (Ravindran and Wadhwa, 2009) developed a probabilistic model to represent the connection between multiple sourcing and its consequences, such as number of bids, the seller's profit, and the buyer's price. Benton (Ravindran and Wadhwa,

2009) developed a nonlinear programming model and a heuristic procedure using Lagrangian relaxation for supplier selection under conditions of multiple items, multiple suppliers, resource limitations, and quantity discounts. Chaudhry et al. (Ravindran and Wadhwa, 2009) developed linear and mixed integer programming models for supplier selection. In their model, price, delivery, quality, and quantity discount were included. Papers by Degraeve (Ravindran and Wadhwa, 2009) and Ghodsypour and O'Brien (Ravindran and Wadhwa, 2009) tackled the supplier selection issue in the framework of TCO or total cost of logistics. Jayaraman et al. (Ravindran and Wadhwa, 2009) formulated a mixed integer linear programming model for solving the supplier selection problem with multiple products, buyers, and suppliers. Feng et al. (Ravindran and Wadhwa, 2009) presented a stochastic integer-programming model for simultaneous selection of tolerances and suppliers based on the quality loss function and process capability index.

Multi-criteria Models

Among the different multi-criteria approaches for supplier selection, GP is the most commonly used method. Buffa and Jackson (Ravindran and Wadhwa, 2009) presented a multi-criteria linear GP model. In this model, two sets of factors are considered: supplier attributes, which include quality, price, service experience, early, late, and on-time deliveries, and the buying firm's specifications, including material requirement and safety stock. Sharma et al. (Ravindran and Wadhwa, 2009) proposed a GP formulation for attaining goals pertaining to price, quality, and lead time under demand and budget constraints. Liu et al. (Ravindran and Wadhwa, 2009) developed a decision support system by integrating the AHP with linear programming. Weber et al. (Ravindran and Wadhwa, 2009) used multi-objective linear programming for supplier selection to systematically analyze the trade-off between conflicting factors. In this model aggregate price, quality and late delivery were considered as goals. Karpak et al. (Ravindran and Wadhwa, 2009) used visual interactive GP for supplier selection process. The objective was to identify and allocate order quantities among suppliers while minimizing product acquisition cost and maximizing quality and reliability. Bhutta and Huq (Ravindran and Wadhwa, 2009) illustrated and compared the technique of TCO and the AHP in supplier selection process. Wadhwa and Ravindran (Ravindran and Wadhwa, 2009) formulated a supplier selection problem with price, lead-time, and quality as three conflicting objectives. The suppliers offered quantity discounts and the model was solved using GP, compromise programming, and weighted objective methods.

MULTI-CRITERIA RANKING METHODS FOR SUPPLIER SELECTION

Many organizations have a large pool of suppliers to select from. The supplier selection problem can be solved in two phases. The first phase reduces the large number of candidate suppliers to a manageable size. In Phase 2, a multiple criteria optimization model is used to allocate order quantities among the shortlisted suppliers.

PRE-QUALIFICATION OF SUPPLIERS

Pre-qualification is defined as the process of reducing the large set of suppliers to a smaller manageable number by ranking the suppliers under a predefined set of criteria. The primary benefits of pre-qualification of suppliers are (Ravindran and Wadhwa, 2009):

1. The possibility of rejecting good suppliers at an early stage is reduced.
2. Resource commitment of the buyer toward purchasing process is optimized.
3. With the application of pre-selected criteria, the pre-qualification process is rationalized.

In this section, we present multiple criteria ranking approaches for the supplier selection problem, namely, the pre-qualification of suppliers.

In the pre-qualification process (Phase 1), readily available qualitative and quantitative data are collected for the various suppliers. This data can be obtained from trade journals, internet, and past transactions to name a few. Once this data is gathered, these suppliers are evaluated using multiple criteria ranking methods. The decision-maker then selects a portion of the suppliers for extensive evaluation in Phase 2.

The first step in pre-qualification is defining the selection criteria. We have used the following 14 pre-qualification criteria as an illustration. The pre-qualification criteria have been split into various subcategories such as Organizational Criteria, Experience criteria, etc. The various pre-qualification criteria are described below:

- Organizational Criteria:
  - Size of company (C1): Size of the company can be either its number of employees or its market capitalization.
  - Age of company (C2): Age of the company is the number of years that the company has been in business.
  - R&D activities (C3): Investment in research and development.
- Experience Criteria:
  - Project type (C4): Specific types of projects completed in the past.
  - Project size (C5): Specific sizes of projects completed in the past.
- Delivery Criteria:
  - Accuracy (C6): Meeting the promised delivery time.
  - Capacity (C7): Capacity of the supplier to fulfill orders.
  - Lead-time (C8): Supplier's promised delivery lead-time.
- Quality Criteria:
  - Responsiveness (C9): If there is an issue concerning quality, how fast the supplier reacts to correct the problem.
  - Defective rate (C10): Rate of defective items among orders shipped.
  - Cost criteria:

- Order change and cancellation charges (C11): Fee associated with modifying or changing orders after they have been placed.
- Unit Cost: Price per item (C12).
- Miscellaneous Criteria:
  - Labor relations (C13): Number of strikes or any other labor problems encountered in the past.
  - Procedural Compliances (C14): Conformance to national/international standards (e.g., ISO 9000).

For the sake of illustration, we assume there are 20 suppliers during pre-qualification. The 14 supplier criteria values for the initial set of 20 suppliers are presented by (Ravindran and Wadhwa, 2009). For all the supplier criteria, larger values are preferred. Next, we discuss several multiple criteria ranking methods for shortlisting the suppliers. Each method has its advantages and limitations. The methods that we discuss here are:

1. The Lp metric method
2. Rating Method
3. Borda Count
4. The Analytic Hierarchy Process (AHP)
5. Cluster Analysis

For a more detailed discussion of multi-criteria ranking methods, the reader is referred to Ravindran (Ravindran and Wadhwa, 2009).

USE OF THE Lp METRIC FOR RANKING SUPPLIERS

Mathematically, the Lp metric represents the distance between two vectors $\mathbf{x}$ and $\mathbf{y}$, where $\mathbf{x}, \mathbf{y} \in R^n$, and is given by:

$$\mathbf{x} - \mathbf{y}_p = \left[\sum_{j=1}^{n} |x_j - y_j|^p \right]^{1/p}$$

One of the most commonly used Lp metrics is the L_2 metric ($p = 2$), which measures the Euclidean distance between two vectors. The ranking of suppliers is done by calculating the L_p metric between the Ideal solution (H) and each vector representing the supplier's ratings for the criteria. The Ideal solution represents the best values possible for each criterion from the initial list of suppliers. Since no supplier will have the best values for all criteria (e.g., a supplier with minimum cost may have poor quality and delivery time), the ideal solution is an artificial target and cannot be achieved. The L_p metric approach computes the distance of each supplier's attributes from the ideal solution and ranks the suppliers based on that distance (smaller the better).

RATING (SCORING) METHOD

Rating is one of the simplest and most widely used ranking methods under conflicting criteria. First, an appropriate rating scale is agreed to (e.g., from 1 to 10, where 10 is the most important and 1 is the least important selection criteria). The scale should be clearly understood to be used properly. Next, using the selected scale, the Decision Maker (DM) provides a rating r_j for each criterion, C_j. The same rating can be given to more than one criterion. The ratings are then normalized to determine the weights of the criteria j. Assuming n criteria:

$$W_j = \frac{r_j}{\sum_{j=1}^{j=n} r_j} \quad \text{for } j = 1, 2, \ldots \ldots n$$

$$Note \sum_{j=1}^{n} w_j = 1$$

Next, a weighted score of the attributes is calculated for each supplier as follows:

$$S_k = \sum_{j=1}^{n} W_j f_{jk} \quad \text{for } k = 1, \ldots K$$

where f_{jk}s are the criteria values for supplier k. The suppliers are then ranked based on their scores. The supplier with the highest score is ranked first. Rating method requires relatively little cognitive burden on the DM.

BORDA COUNT

This method is named after Jean Charles de Borda, the eighteenth century French Physicist. The method is as follows:

- The n criteria are ranked 1(most important) to n (least important)
  - Criterion ranked 1 gets n points, criterion ranked 2 gets $n-1$ points, and the last criterion gets 1 point.
- Weights for the criteria are calculated as follows:
  - Criterion ranked 1 = n/S
  - Criterion ranked 2 = $(n-1)/S$
  - last criterion = $1/S$
    where S is the sum of all the points = $n(n+1)/2$.

ANALYTIC HIERARCHY PROCESS (AHP)

The AHP, developed by Saaty (Ravindran and Wadhwa, 2009), is a MCDM method for ranking alternatives. Using the AHP, the decision-maker can assess not only quantitative but also various intangible factors such as financial stability, feeling of trust, etc., in the supplier selection process. The buyer establishes a set of evaluation criteria and the AHP uses these criteria to rank the different suppliers. The AHP can enable the DM to represent the interaction of multiple factors in complex and unstructured situations.

Basic Principles of the AHP

- Design a Hierarchy: Top vertex is the main objective and bottom vertices are the alternatives. Intermediate vertices are criteria/sub-criteria (which are more and more aggregated as you go up in the hierarchy).
- At each level of the hierarchy, a paired comparison of the vertices criteria/ sub-criteria is performed from the point of view of their "contribution (weights)" to each of the higher-level vertices to which they are linked.
- Uses both rating method and comparison method. A numerical scale 1–9 (1-equal importance; 9-most important).
- Uses pairwise comparison of alternatives with respect to each criterion (sub-criterion) and gets a numerical score for each alternative on every criterion (sub-criterion).
- Compute total weighted score for each alternative and rank the alternatives accordingly.

Steps of the AHP Model

Step 1: In the first step, carry out a pairwise comparison of criteria using the 1–9 degree of importance scale (see Ravindran and Wadhwa, 2009).

If there are n criteria to evaluate, then the pairwise comparison matrix for the criteria is given by; $A_{(NxN)} = [a_{ij}]$, where a_{ij} represents the relative importance of criterion i with respect to criterion j. Set $a_{ii} = 1$ and $a_{ji} = \dfrac{1}{a_{ij}}$.

Step 2: Compute the normalized weights for the main criteria. We obtain the weights using L_1 norm. The two-step process for calculating the weights is as follows:

- Normalize each column of $A_{(NxN)}$ using L_1 norm.

$$r_{ij} = \frac{a_{ij}}{\sum\limits_{i=1}^{n} a_{ij}}$$

- Average the normalized values across each row.

$$w_i = \frac{\sum\limits_{j=1}^{n} r_{ij}}{n}$$

Step 3: In the third step, we check for consistency of the pairwise comparison matrix using the eigen value theory as follows (Ravindran and Wadhwa, 2009):

1. Using the pairwise comparison matrix A and the weights W compute AW. Let the vector $X = (X_1, X_2, X_3 .... X_n)$ denote the values of AW.
2. Compute

$$\lambda_{max} = \text{Average}\left[\frac{X_1}{W_1}, \frac{X_2}{W_2}, \frac{X_3}{W_3} ... \frac{X_n}{W_n}\right]$$

3. Consistency Index (CI) is given by

$$CI = \frac{\lambda_{max} - n}{n - 1}$$

Saaty (Ravindran and Wadhwa, 2009) generated a number of random positive reciprocal matrices with $a_{ij} \in (1, 9)$ for different sizes and computed their average CI values, denoted by RI, as given below. He defines the Consistency Ratio (CR) as $CR = \frac{CI}{RI}$. If $CR < 0.15$, then accept the pairwise comparison matrix as consistent.

Step 4: In the next step, we compute the relative importance of the sub-criteria in the same way as done for the main criteria. Step 2 and Step 3 are carried out for every pair of sub-criteria with respect to their main criterion. The final weights of the sub-criteria are the product of the weights along the corresponding branch.

Step 5: Repeat Steps 1, 2, and 3 and obtain,
   a. Pair-wise comparison of alternatives with respect to each criterion using the ratio scale (1–9).
   b. Normalized scores of all alternatives with respect to each criterion. Here, an (mxm) matrix S is obtained, where S_{ij} = normalized score for alternative "" with respect to criterion "j" and m is the number of alternatives.

Step 6: Compute the total score (TS) for each alternative as follows $TS_{(mx1)} = S_{(mxn)} W_{(nx1)}$, where W is the weight vector obtained after Steps 3 and 4. Using the TSs, the alternatives are ranked. There is commercially available software for the AHP called Expert Choice.

CLUSTER ANALYSIS

Clustering analysis (CA) is a statistical technique particularly suited to grouping of data. It is gaining wide acceptance in many different fields of research such as data mining, marketing, operations research, and bioinformatics. CA is used when it is believed that the sample units come from an unknown population. Clustering is the classification of similar objects into different groups, or more precisely, the partitioning of a data set into subsets (clusters), so that the data in each subset share some common trait. CA develops subsets of the raw data such that each subset contains member of like nature (similar supplier characteristics) and that difference between different subsets is as pronounced as possible.

There are two types of clustering algorithms (Ravindran and Wadhwa, 2009), namely

- **Hierarchical**: Algorithms which employ hierarchical clustering find successive clusters using previously established clusters. Hierarchical algorithms can be further classified as *agglomerative* or *divisive*. Agglomerative algorithms begin with each member as a separate cluster and merge them into successively larger clusters. On the other hand, divisive algorithms begin with the whole set as one cluster and proceed to divide it into successively smaller clusters. Agglomerative method is the most common hierarchical method.
- **Partitional**: In partitional clustering, the algorithm finds all the clusters at once. An example of partitional methods is k-means clustering.

PROCEDURE FOR CLUSTER ANALYSIS

Clustering process begins with formulating the problem and concludes with carrying out analysis to verify the accuracy and appropriateness of the method. The clustering process has the following steps:

1. Formulate the problem and identify the selection criteria.
2. Decide on the number of clusters.
3. Select a clustering procedure.
4. Plot the dendrogram (A dendrogram is a tree diagram used to illustrate the output of CA) and carry out analysis to compare the means across various clusters.

Let us illustrate the cluster analysis using our supplier selection example.

Step 1: In the first step, every supplier is rated on a scale of 0–1 for each attribute.

Step 2: In this step, we need to decide on the number of clusters. We want the initial list of suppliers to be split into two categories, good suppliers and bad suppliers; hence, the number of clusters is two.

Step 3: Next, we apply both hierarchical and partitional clustering methods to supplier data. We choose the method which has the highest R-sq value pooled over all the 14 attributes.

The R-sq value for k-means is the highest among different methods; hence, k-means is chosen for clustering. There are several other methods available for determining the goodness of fit (Ravindran and Wadhwa, 2009).

COMPARISON OF RANKING METHODS

Different ranking methods can provide different solutions resulting in rank reversals. In extensive empirical studies with human subjects, it has been found (Ravindran and Wadhwa, 2009) that Borda count (with pairwise comparison of criteria) rankings are generally in line with the AHP rankings. Given the increased cognitive burden and expensive calculations required for the AHP, Borda count might be selected as an appropriate method for supplier rankings. Even though the Rating method is easy to use, it could lead to several ties in the final rankings, thereby making the results less useful.

GROUP DECISION-MAKING

Most purchasing decisions, including the ranking and selection of suppliers, involve the participation of multiple DMs, and the ultimate decision is based on the aggregation of the DM's individual judgments to arrive at a group decision. The rating method, Borda count, and the AHP discussed in this section can be extended to group decision-making as described below:

1. Rating Method: Ratings of each DM for every criterion is averaged. The average ratings are then normalized to obtain the group criteria weights.
2. Borda Count: Points are assigned based on the number of DMs that assign a particular rank for a criterion. These points are then totaled for each criterion and normalized to get criteria weights. (This is similar to how the college polls are done to get the top 25 football or basketball teams.)
3. The AHP: There are two methods to get the group rankings using the AHP.
   a. Method 1: Strength of preference scores assigned by the individual DMs are aggregated using geometric means and then used in the AHP calculations.
   b. Method 2: First, all the alternatives are ranked by each DM using the AHP. The individual rankings are then aggregated to a group ranking using Borda count.

MULTI-OBJECTIVE SUPPLIER ALLOCATION MODEL

As a result of pre-qualification, the large number of initial suppliers is reduced to a manageable size. In the second phase of the supplier selection, detailed quantitative data such as price, capacity, quality, etc., are collected on the shortlisted suppliers and are used in a multi-objective framework for the actual order allocation. We consider multiple buyers, multiple products, and multiple suppliers with volume discounts. The scenario of multiple buyers is possible in case of a central purchasing department, where different divisions of an organization buy through one purchasing department. Here, the number of buyers will be equal to the number of divisions buying through the central purchasing. In all other cases, the number of buyers is equal to one. We consider the least restrictive case where any of the buyers can acquire one or more products from any suppliers, namely, a multiple sourcing model.

In this phase of the supplier selection process, an organization will make the following decisions:

- To choose the most favorable suppliers who would meet its supplier selection criteria for the various components.
- To order optimal quantities from the chosen most favorable supplier to meet its production plan or demand.

The mathematical model for the order allocation problem is discussed next.

NOTATIONS USED IN THE MODEL

Model Indices

I: Set of products to be purchased

J: Set of buyers who procure multiple units in order to fulfill some demand

K: Potential set of suppliers

M: Set of incremental price breaks for volume discounts

Model Parameters

p_{ikm}: Cost of acquiring one unit of product i from supplier k at price level m.

b_{ikm}: Quantity at which incremental price breaks occurs for product i by supplier k.

F_k: Fixed ordering cost associated with supplier k.

d_{ij}: Demand of product i for buyer j.

l_{ijk}: Lead time of supplier k to produce and supply product i to buyer j. The lead time of different buyers could be different because of geographical distances.

q_{ik}: Quality that supplier k maintains for product i, which is measured as percent of defects.

CAP_{ik}: Production capacity for supplier k for product i.

N: Maximum number of suppliers that can be selected.

Decision Variables in the Model

X_{ijkm}: Number of units of product i supplied by supplier k to buyer j at price level m.

Z_k: Denotes if a particular supplier is chosen or not. This is a binary variable which takes a value 1 if a supplier is chosen to supply any product and is 0, if the supplier is not chosen at all.

Y_{ijkm}: This is binary variable which takes on a value 1 if price level m is used, 0 otherwise.

Mathematical Formulation of the Order Allocation Problem

The conflicting objectives used in the model are simultaneous minimization of price, lead-time, and rejects. It is relatively easy to include other objectives also. The mathematical form for these objectives is:

1. Price (z_1): Total cost of purchasing has two components; fixed and the variable cost.

   Total variable cost: The total variable cost is the cost of buying every additional unit from the suppliers and is given by:

$$\sum_i \sum_j \sum_k \sum_m p_{ikm} \cdot X_{ijkm}$$

   Fixed Cost: If a supplier k is used then there is a fixed cost associated with it, which is given by:

$$\sum_k F_k \cdot Z_k$$

   Hence, the total purchasing cost is,

$$\sum_i \sum_j \sum_k \sum_m p_{ikm} \cdot X_{ijkm} + \sum_k F_k \cdot Z_k$$

2. Lead-time (z_2): $\sum_i \sum_j \sum_k \sum_m l_{ijk} \cdot X_{ijkm}$

   The product of lead-time of each product and quantity supplied is summed over all the products, buyers, and suppliers and should be minimized.

3. Quality (z_3): $\sum_i \sum_j \sum_k \sum_m q_{ik} \cdot X_{ijkm}$

The product of rejects and quantity supplied is summed over all the products, buyers, and suppliers and should be minimized. Quality in our case is measured in terms of percentage of rejects.

The constraints in the model are as follows:

1. **Capacity constraint**: Each supplier k has a maximum capacity for product i, CAP_{ik}. Total order placed with this supplier must be less than or equal to the maximum capacity. Hence the capacity constraint is given by:

$$\sum_i \sum_j \sum_m X_{ijkm} \leq \left(CAP_{ik}\right) Z_k \quad \forall k$$

   The binary variable on the right-hand side of the constraint implies that a supplier cannot supply any products if not chosen, i.e., if Z_k is 0.

2. **Demand Constraint**: The demand of buyer j for product i has to be satisfied using a combination of the suppliers. The demand constraint is given by:

$$\sum_k \sum_m X_{ijkm} = d_{ij} \quad \forall i, j$$

3. **Maximum number of suppliers**: The maximum number of suppliers chosen must be less than or equal to the specified number. Hence this constraint takes the following form:

$$\sum_k Z_k \leq N$$

4. **Linearizing constraints**: In the presence of incremental price discounts, objective function is nonlinear. The following set of constraints are used to linearize it:

$$X_{ijkm} \leq \left(b_{ikm} - b_{ikm-1} \right) * Y_{ijkm} \forall i, j, k, 1 \leq m \leq m_k$$
$$X_{ijkm} \geq \left(b_{ikm} - b_{ikm-1} \right) * Y_{ijkm+1} \forall i, j, k, 1 \leq m \leq m_k - 1$$

$0 = b_{i,k,0} < b_{i,k,1} < \ldots < b_{i,k,mk}$ is the sequence of quantities at which price break occurs. p_{ikm} is the unit price of ordering X_{ijkm} units from supplier k at level m, if $b_{i,k,m-1} < X_{ijkm} \leq b_{i,k,m}$ $(1 \leq m \leq m_k)$.

The above constraints force quantities in the discount range for a supplier to be incremental. Because the "quantity" is incremental, if the order quantity lies in discount interval m, namely, $Y_{ijkm} = 1$, then the quantities in interval 1 to $m-1$, should be at the maximum of those ranges. Constraint (16) also assures that a quantity in any range is no greater than the width of the range.

5. **NonNegativity and Binary constraint**:

$$X_{ijkm} \geq 0 \; ; Z_k, Y_{ijkm} \in (0,1) \tag{6.1}$$

GOAL PROGRAMMING METHODOLOGY

One way to treat multiple criteria is to select one criterion as primary and the other criteria as secondary. The primary criterion is then used as the optimization objective function, while the secondary criteria are assigned acceptable minimum and maximum values and are treated as problem constraints. However, if careful considerations were not given while selecting the acceptable levels, a feasible solution that satisfies all the constraints may not exist. This problem is overcome by *GP*, which has become a popular practical approach for solving multiple criteria optimization problems.

GP falls under the class of methods that use completely pre-specified preferences of the decision-maker in solving the Multi-criteria mathematical programming problems. In GP, all the objectives are assigned target levels for achievement and a relative priority on achieving those levels. GP treats these targets as *goals to aspire for* and not as absolute constraints. It then attempts to find an optimal solution that comes as "close as possible" to the targets in the order of specified priorities. In this section, we discuss how to formulate GP models and their solution methods.

Before we discuss the formulation of GP problems, we discuss the difference between the terms *real constraints* and *goal constraints* (or simply *goals*) as used in GP models. The real constraints are absolute restrictions on the decision variables, while the goals are conditions one would like to achieve but are not mandatory. For instance, a real constraint given by

$$x_1 + x_2 = 3$$

requires all possible values of $x_1 + x_2$ to always equal 3. As opposed to this, a goal requiring $x_1 + x_2 = 3$ is not mandatory, and we can choose values of $x_1 + x_2 \geq 3$ as well as $x_1 + x_2 \leq 3$. In a goal constraint, positive and negative deviational variables are introduced to represent constraint violations as follows:

$$x_1 + x_2 + d_1^- - d_1^+ = 3 d_1^+, d_1^- \geq 0$$

Note that, if $d_1^- > 0$, then $x_1 + x_2 \langle 3,$ and if $d_1^+ \rangle 0,$ then $x_1 + x_2 > 3.$

By assigning suitable weights w_1^- and w_1^+ on d_1^- and d_1^+ in the objective function, the model will try to achieve the sum $x_1 + x_2$ as close as possible to 3. If the goal were to satisfy $x_1 + x_2 \geq 3$, then only d_1^- is assigned a positive weight in the objective, while the weight on d_1^+ is set to 0.

GENERAL GOAL PROGRAMMING MODEL

A general Multiple Criteria Mathematical Programming (MCMP) problem is given as follow:

$$\max \mathbf{F}(\mathbf{x}) = \left\{ f_1(\mathbf{x}), f_2(\mathbf{x}), \ldots, f_k(\mathbf{x}) \right\}$$

Subject to

$$g_j(\mathbf{x}) \leq 0 \text{ for } j = 1, \ldots, m$$

where $\mathbf{x}$ is an n-vector of *decision variables*, and $f_i(\mathbf{x})$, $i = 1, \ldots, k$ are the k *criteria/objective functions*.

Let $S = \{ \mathbf{x}/g_j(\mathbf{x}) \leq 0, \text{ for all "j"} \}$

$Y = \{ \mathbf{y}/\mathbf{F}(\mathbf{x}) = \mathbf{y} \text{ for some } \mathbf{x} \in S \}$

S is called the *Decision space* and Y is called the *criteria or objective space* in MCMP.

Consider the general MCMP problem presented earlier. The assumption that there exists an optimal solution to the MCMP problem involving multiple criteria implies the existence of some preference ordering of the criteria by the DM. The GP formulation of the MCMP problem requires the DM to specify an acceptable level of achievement (b_i) for each criterion f_i and specify a weight w_i (ordinal or cardinal) to be associated with the deviation between f_i and b_i. Thus, the GP model of an MCMP problem becomes:

$$\text{Minimize } Z = \sum_{i=1}^{k} \left(w_i^+ d_i^+ + w_i d_i^- \right)$$

Subject to:

$$f_i(x) + d_i^- - d_i^+ = b_i \text{ for } i = 1, \ldots, k$$

$$g_j(x) \leq 0 \text{ for } j = 1, \ldots, m$$

$$x_j, d_i, d_i^+ \geq 0 \text{ for all } i \text{ and } j$$

The equation above represents the objective function of the GP model, which minimizes the weighted sum of the deviational variables. The system of equations represents the goal constraints relating the multiple criteria to the goals/targets for those criteria. The variables, d_i^- and d_i^+, are the deviational variables, representing the under achievement and over achievement of the ith goal. The set of weights (w_i^+ and w_i^-) may take two forms:

1. Pre-specified weights (cardinal).
2. Preemptive priorities (ordinal).

Under pre-specified (cardinal) weights, specific values in a relative scale are assigned to w_i^+ and w_i^- representing the DM's "trade-off" among the goals. Once w_i^+ and w_i^- are specified, the goal program reduces to a single objective optimization problem. The cardinal weights could be obtained from the DM using any of the methods discussed earlier (Rating, Borda count, and the AHP). However, for this method to work effectively, criteria values have to be scaled properly. In reality, goals are usually incompatible (i.e., in commensurable) and some goals can be achieved only at the expense of some other goals. Hence, preemptive GP, which is more common in practice, uses ordinal ranking or preemptive priorities to the goals by assigning incommensurable goals to different priority levels and weights to goals at the same priority level. In this case, the objective function of the GP model takes the form:

$$\text{Minimize } Z = \sum_{p} P_p \sum_{i} \left(w_{ip}^+ d_i^+ + w_{ip}^- d_i^- \right)$$

where P_p represents priority p with the assumption that P_p is much larger then P_{p+1} and w_{ip}^+ and w_{ip}^- are the weights assigned to the ith deviational variables at priority p. In this manner, lower priority goals are considered only after attaining the higher priority goals. Thus, preemptive GP is essentially a sequential single objective optimization process, in which successive optimizations are carried out on the alternate optimal solutions of the previously optimized goals at higher priority. In addition to preemptive and nonpreemptive GP models, other approaches (Fuzzy GP, Min-Max GP) have also been proposed.

PREEMPTIVE GOAL PROGRAMMING

For the three criteria supplier order allocation problems, the preemptive GP formulation will be as follows:

$$\min P_1 d_1^+ + P_2 d_2^+ + P_3 d_3^+ \tag{6.2}$$

Subject to.

$$\sum_i \sum_j \sum_k \sum_m l_{ijk} \cdot x_{ijkm} + d_1^- - d_1^+ = \text{Lead time goal} \tag{6.3}$$

$$\left(\sum_i \sum_j \sum_k \sum_m p_{ikm} \cdot x_{ijkm} + \sum_k F_k \cdot z_k \right) + d_2^- - d_2^+ = \text{Price goal} \tag{6.4}$$

$$\sum_i \sum_j \sum_k \sum_m q_{ik} \cdot x_{ijkm} + d_3^- - d_3^+ = \text{Quality goal} \tag{6.5}$$

$$d_n^-, d_n^+ \geq 0 \quad \forall n \in \{1,\dots,3\} \tag{6.6}$$

$$\sum_j \sum_m x_{ijkm} \leq CAP_{ik} \cdot z_k \quad \forall i,k \tag{6.7}$$

$$\sum_k \sum_m x_{ijkm} = D_{ij} \quad \forall i,j \tag{6.8}$$

$$\sum_k z_k \leq N \tag{6.9}$$

$$x_{ijkm} \leq \left(b_{ikm} - b_{ik(m-1)} \right) \cdot y_{ijkm} \quad \forall i,j,k \quad 1 \leq m \leq m_k \tag{6.10}$$

$$x_{ijkm} \geq \left(b_{ikm} - b_{ik(m-1)} \right) \cdot y_{ijk(m+1)} \quad \forall i,j,k \quad 1 \leq m \leq m_k - 1 \tag{6.11}$$

$$x_{ijkm} \geq 0 \quad z_k \in \{0,1\} \quad y_{ijkm} \in \{0,1\} \tag{6.12}$$

NONPREEMPTIVE GOAL PROGRAMMING

In nonpreemptive GP model, the buyer sets goals to achieve for each objective and preferences in achieving those goals expressed as numerical weights. In the non-preemptive GP, the buyer has three goals, namely,

- Limit the lead-time to Lead goal with weight w_1.
- Limit the total purchasing cost to Price goal with weight w_2.
- Limit the quality to Quality goal with weight w_3.

The weights w_1, w_2, and w_3 can be obtained using the methods discussed earlier. The non-preemptive GP model can be formulated as

$$\min Z = w_1 * d_1^+ + w_2 * d_2^+ + w_3 * d_3^+$$

subject to the constraints to the earlier constraints.

In the above model, d_1^+, d_2^+, and d_3^+ represent the overachievement of the stated goals. Due to the use of the weights the model needs to be scaled. The weights w_1, w_2, and w_3 can be varied to obtain different GP optimal solutions.

TCHEBYCHEFF (MIN – MAX) GOAL PROGRAMMING

In this GP model, the DM only specifies the goals/targets for each objective. The model minimizes the maximum deviation from the stated goals. For the supplier selection problem, the Tchebycheff goal program becomes:

Min Max (d_1^+, d_2^+, d_3^+)

$$d_i^+ \geq 0 \quad \forall i$$

The above can be reformulated as a linear objective by setting

$$\max \left(d_1^+, d_2^+, d_3^+ \right) = M \geq 0$$

Thus, we have the following equivalent equation:

$$\min Z = M$$

Subject to:

$$M \geq \left(d_1^+\right)$$

$$M \geq \left(d_2^+\right)$$

$$M \geq \left(d_3^+\right)$$

$$d_i^+ \geq 0 \ \forall i$$

The constraints stated earlier will also be included in this model. The advantage of Tchebycheff goal program is that there is no need to get preference information (priorities or weights) about goal achievements from the DM. Moreover, the problem reduces to a single objective optimization problem. The disadvantages of this method are (i) the scaling of goals is necessary (as required in nonpreemptive GP) and (ii) outliers are given more importance and could produce poor solutions.

FUZZY GOAL PROGRAMMING

Fuzzy GP uses the ideal values as targets and minimizes the maximum normalized distance from the ideal solution for each objective. An ideal solution is the vector of best values of each criterion obtained by optimizing each criterion independently ignoring other criteria. In this example, ideal solution is obtained by minimizing price, lead-time, and quality independently. In most situations, ideal solution is an infeasible solution since the criteria conflict with one another.

If M equals the maximum deviation from the ideal solution, then the Fuzzy GP model is as follows:

$$\min Z = M$$

Subject to:

$$M \geq \left(d_1^+\right) / \lambda_1$$

$$M \geq \left(d_2^+\right) / \lambda_2$$

$$M \geq \left(d_3^+\right) / \lambda_3$$

$$d_i^+ \geq 0 \ \forall i$$

The constraints stated earlier will also be included in this model. In the above model λ_1, λ_2, and λ_3 are scaling constants to be set by the user. A common practice is to set the values λ_1, λ_2, λ_3 equal to the respective ideal values. The advantage of Fuzzy GP is that no target values have to be specified by the DM.

For additional readings on the variants of Fuzzy GP models, the reader is referred to Ignizio and Cavalier (Ravindran and Wadhwa, 2009), Tiwari et al. (Ravindran and Wadhwa, 2009), Mohammed (Ravindran and Wadhwa, 2009), and Hu et al. (Ravindran and Wadhwa, 2009).

An excellent source of reference for GP methods and applications is the textbook by Schniederjans (Ravindran and Wadhwa, 2009).

See (Ravindran and Wadhwa, 2009) for a case study that illustrates the four GP methods using a supplier order allocation.

CONCLUDING REMARKS

While supplier selection plays an important role in purchasing, it is especially important for cases of supply disruption, such as the COVID-19 pandemic. This chapter illustrates the use of both discrete and continuous MCDM techniques to optimize the supplier selection process. In this chapter, we present the supplier selection problem in two phases. In the first phase called, pre-qualification, we reduce the initial set of large number suppliers to a manageable set. Phase 1 reduces the effort of the buyer and makes the pre-qualification process entirely objective. In the second phase, we analyze the shortlisted suppliers using the multi-objective technique known as GP. We consider several conflicting criteria, including, price, lead-time, and quality. An important distinction of multi-objective techniques is that it does not provide one optimal solution but a number of solutions known as efficient solutions. Hence, the role of the decision-maker (buyer) is more important than before. By involving the decision-maker early in the process, the acceptance of the model results by management becomes easier. The efficient solutions are compared using the Value Path Approach to show the criteria trade-off obtained using different GP approaches. Besides GP, there are other approaches to solve the multi-criteria optimization model for the supplier selection problem. They include Weighted Objective Method, Compromise Programming, and Interactive Approaches. Interested readers can refer to Wadhwa and Ravindran (Ravindran and Wadhwa, 2009) and Ravindran (Ravindran and Wadhwa, 2009) for more details. Reference (Ravindran and Wadhwa, 2009) also provides information on the computer software available for MCDM methods.

The supplier selection models can be extended in many different directions. One of the areas is managing supplier risk. Along with cost, quality, technology, and service criteria used in sourcing decisions, there is a need to integrate global risk factors such as political stability, currency fluctuations, taxes, local content rules, infrastructure (ports, raw materials, communication, transportation, certification, pandemic lockdown, etc.) in the supplier selection process. Another area of research is *supplier monitoring*. Since the supplier performance factors can change over time, real-time monitoring of suppliers becomes critical. The monitoring issues are to determine what supplier performance factors to monitor and the frequency of monitoring.

REFERENCES

Ravindran, A. R. (2008), *Operations Research and Management Science Handbook*, CRC Press, Boca Raton, FL.

Ravindran, A. R. and Vijay Wadhwa (2009), "Multiple Criteria Optimization Models for Supplier Selection," in Badiru, A. B. and M. Thomas, editors (2009), *Handbook of Military Industrial Engineering*, CRC Press/Taylor & Francis Group, Boca Raton, FL.

Ravindran, A. R., Paul M. Griffin, and Vittaldas V. Prabhu (2018), *Service Systems Engineering and Management*, CRC Press/Taylor & Francis Group, Boca Raton, FL.

Ravindran, A. R. and Donald P. Warsing, Jr. (2017), *Supply Chain Engineering: Models and Applications*, CRC Press/Taylor & Francis Group, Boca Raton, FL.

Wadhwa, V. and Ravindran, A. R. (2007), "Vendor selection in outsourcing," *Computers & Operations Research*, Vol. 34, p. 3725.

7 Systems Models for the Supply Chain

INTRODUCTION TO SYSTEMS ENGINEERING FOR SUPPLY CHAIN

As introduced in Chapter 1, systems engineering (SE) is the application of engineering to solutions of a multifaceted problem through a systematic collection and integration of parts of the problem with respect to the life cycle of the problem. It is the branch of engineering concerned with the development, implementation, and use of large or complex systems. It focuses on specific goals of a system considering the specifications, prevailing constraints, expected services, possible behaviors, and structure of the system. It also involves a consideration of the activities required to assure that the system's performance matches the stated goals. SE addresses the integration of tools, people, and processes required to achieve a cost-effective and timely operation of the system. In this context, a system is defined as a collection of interrelated elements, whose collective output (together) is higher than the sum of the outputs of the individual elements.

Perez (2014) defines supply chain as a business function responsible for the connection and combination of activities related to the management, within and among organizations, of the flow of materials, products, information, and commercial and financial transactions.

The above definition, essentially, describes the supply chain as a system. As such, SE modeling is applicable and appropriate for the supply chain.

The complexity of supply chains in a disruptive pandemic cannot be overstated. The entire world is adversely affected. Manufacturers and retailers are all scrambling to figure out how to respond to the pandemic and move their goods to the points of need. This chapter provides cogent guidelines for business and industry to manage their supply chains using SE models, tools, techniques, and concepts. The methodology of SE combines both qualitative and quantitative tools.

INTRODUCTION TO THE DEJI SYSTEMS MODEL FOR SUPPLY CHAIN

Supply chain is at the intersection of efficiency, effectiveness, and productivity. Efficiency provides the framework for quality in terms of resources and inputs required to achieve the desired level of quality and output of the supply chain.

DOI: 10.1201/9781003111979-7

Effectiveness comes into play with respect to the application of product quality to meet specific needs and requirements of an organization. Productivity is an essential factor in the pursuit of quality as it relates to the throughput of a production system. To achieve the desired levels of quality, efficiency, effectiveness, and productivity, an appropriate supply chain modeling framework must be adopted. This chapter presents a supply chain enhancement methodology based on a systems approach for the Design, Evaluation, Justification, and Integration (DEJI) model. The model is relevant for research efforts in quality engineering and technology applications and other SE applications.

The DEJI model of SE provides one additional option for SE development applications. Although the model is generally applicable in all types of systems modeling, systems quality is specifically used to describe how the DEJI systems model can be applied to the supply chain. The core stages of the DEJI model are:

- Design
- Evaluation
- Justification
- Integration

Design encompasses any system initiative providing a starting point for a project. Thus, design can include technical product design, process initiation, and concept development. In essence, we can say that "design" represents requirements and specifications. Evaluation can use a variety of metrics both qualitative and quantitative, depending on the organization's needs. Justification can be done on the basis of monetary, technical, or social reasons. Integration needs to be done with respect to the normal or standard operations of the organization. Figure 7.1 illustrates the full profile of the DEJI systems model®, which is trademarked as a SE tool.

FIGURE 7.1   The DEJI Systems Model.

All the operational elements embedded in the DEJI model are described below:

- Design embodies Agility, Define End Goal, and Engage Stakeholder.
- Evaluate embodies Feasibility, Metrics, Gather Evidence, and Assess Utility.
- Justify embodies Desirability, Focus on Implementation, and Articulate Conclusions.
- Integrate embodies Affordability, Sustainability, and Practicality.

For application purposes, these elements interface and interact systematically to enhance overall operational performance of an organization.

Several aspects of quality in the supply chain must undergo rigorous research along the realms of both quantitative and qualitative characteristics. Many times, quality is taken for granted and the flaws only come out during the implementation stage, which may be too late to rectify. The growing trend in product recalls is a symptom of a priori analysis of the sources and implications of quality at the product conception stage. This chapter advocates the use of the DEJI systems model for enhancing quality design, quality evaluation, quality justification, and quality integration through hierarchical and stage-by-stage processes in the supply chain.

Better quality is achievable and there is always room for improvement in the quality of products and services. But we must commit more efforts to the research at the outset of the supply chain design cycle. Even the human elements of the perception of quality can benefit from more directed research from a social and behavioral sciences point of view.

QUALITY ACCOUNTABILITY

Throughout history, engineering has answered the call of the society to address specific challenges (Badiru, 2019). With such answers come a greater expectation of professional accountability. Consider the level of social responsibility that existed during the time of the Code of Hammurabi. Two of the laws are presented below:

> Hammurabi's Law 229: "If a builder builds a house for someone, and does not construct it properly, and the house which he built fall in and kill its owner, then that builder shall be put to death."
>
> Hammurabi's Law 230: "If it kills the son of the owner the son of that builder shall be put to death."

These are drastic measures designed to curb professional dereliction of duty and enforce social responsibility with particular focus on product quality, particularly where a supply chain is involved. Research and education must play bigger and more direct roles in the design, practice, and management of quality. With the above principles as possible tenets for better research, education, and practice of quality in engineering and technology, this chapter presents the DEJI systems model as a potential methodology for enhancing a systems approach to the supply chain.

The DEJI model encourages the practice of building quality into a product right from the beginning so that the product integration stage can be more successful.

DESIGN OF QUALITY

The design of quality in product development should be structured to follow point-to-point transformations. A good technique to accomplish this is the use of state-space transformation, with which we can track the evolution of a product from the concept stage to a final product stage. For the purpose of product quality design, the following definitions are applicable:

Product state: A state is a set of conditions that describe the product at a specified point in time. The *state* of a product refers to a performance characteristic of the product which relates input to output such that a knowledge of the input function over time and the state of the product at time $t = t_0$ determines the expected output for $t \geq t_0$. This is particularly important for assessing where the product stands in the context of new technological developments and the prevailing operating environment.

Product state-space: A product *state-space* is the set of all possible states of the product life cycle. State-space representation can solve product design problems by moving from an initial state to another state, and eventually to the desired end-goal state. The movement from state to state is achieved by means of actions. A goal is a description of an intended state that has not yet been achieved. The process of solving a product problem involves finding a sequence of actions that represents a solution path from the initial state to the goal state. A state-space model consists of state variables that describe the prevailing condition of the product. The state variables are related to inputs by mathematical relationships. Examples of potential product state variables include schedule, output quality, cost, due date, resource, resource utilization, operational efficiency, productivity throughput, and technology alignment. For a product described by a system of components, the state-space representation can follow the quantitative metric below:

$$Z = f(z, x); \quad Y = g(z, x)$$

where f and g are vector-valued functions. The variable Y is the output vector while the variable x denotes the inputs. The state vector Z is an intermediate vector relating x to y. In generic terms, a product is transformed from one state to another by a driving function that produces a transitional relationship given by:

$$S_s = f\left(x|S_p\right) + e,$$

where S_s = subsequent state; x = state variable; S_p = the preceding state; e = error component

The function f is composed of a given action (or a set of actions) applied to the product. Each intermediate state may represent a significant milestone in the project. Thus, a descriptive state-space model facilitates an analysis of what actions to apply in order to achieve the next desired product state. A systems linkage will show a representation of a product development involving the transformation of a product from one state to another through the application of human or machine actions, in a supply chain. This simple structure can be expanded to cover several components within the product information framework. Hierarchical linking of product elements provides an expanded transformation structure. The product state can be expanded in accordance with implicit requirements. These requirements might include grouping of design elements, linking precedence requirements (both technical and procedural), adapting to new technology developments, following required communication links, and accomplishing reporting requirements. The actions to be taken at each state depend on the prevailing product conditions, just as in supply chain network connections. The nature of subsequent alternate states depends on what actions are implemented. Sometimes there are multiple paths that can lead to the desired end result. At other times, there exists only one unique path to the desired objective. In conventional practice, the characteristics of the future states can only be recognized after the fact, thus, making it impossible to develop adaptive plans. In the implementation of the DEJI systems model, adaptive plans can be achieved because the events occurring within and outside the product state boundaries can be taken into account. If we describe a product by P state variables s_i, then the composite state of the product at any given time can be represented by a vector $\mathbf{S}$ containing P elements. That is,

$$\mathbf{S} = \left\{s_1, s_2, \ldots, s_P\right\}$$

The components of the state vector could represent either quantitative or qualitative variables (e.g., cost, energy, color, and time). We can visualize every state vector as a point in the state-space of the product. The representation is unique since every state vector corresponds to one and only one point in the state-space. Suppose we have a set of actions (transformation agents) that we can apply to the product information so as to change it from one state to another within the project state-space. The transformation will change a state vector into another state vector. A transformation may be a change in raw material or a change in design approach. The number of transformations available for a product characteristic may be finite or unlimited. We can construct trajectories that describe the potential states of a product evolution as we apply successive transformations with respect to technology forecasts. Each transformation may be repeated as many times as needed. Given an initial state S_0, the sequence of state vectors is represented by the following:

$$S_n = T_n\left(S_{n-1}\right),$$

The state-by-state transformations are then represented as $S_1 = T_1(S_0)$; $S_2 = T_2(S_1)$; $S_3 = T_3(S_2)$; ...; $S_n = T_n(S_{n-1})$. The final state, S_n, depends on the initial state S and the effects of the actions applied.

EVALUATION OF SUPPLY CHAIN QUALITY

The second stage of the DEJI systems model is evaluation. A product can be evaluated on the basis of cost, quality, schedule, and meeting requirements. There are many quantitative metrics that can be used in evaluating a product at this stage. Learning curve productivity is one relevant technique that can be used because it offers an evaluation basis of a product with respect to the concept of growth and decay. The half-life extension (Badiru, 2012) of the basic learning is directly applicable because the half-life of the technologies going into a product can be considered. In today's technology-based operations, retention of learning may be threatened by fast-paced shifts in operating requirements. Thus, it is of interest to evaluate the half-life properties of new technologies as they impact the overall product quality. Information about the half-life can tell us something about the sustainability of learning-induced technology performance. This is particularly useful for designing products whose life cycles stretch into the future in a high-tech environment.

JUSTIFICATION OF QUALITY

The third phase of the DEJI systems model is justification. We need to justify a program on the basis of quantitative value assessment. The Systems Value Model (SVM) is a good quantitative technique that can be used here for project justification on the basis of value. The model provides a heuristic decision aid for comparing project alternatives. It is presented here again for the present context. Value is represented as a deterministic vector function that indicates the value of tangible and intangible attributes that characterize the project. It is represented as follows:

$$V = f\left(A_1, A_2, \ldots, A_p\right),$$

where V is the assessed value and the A values are quantitative measures or attributes. Examples of product attributes are quality, throughput, manufacturability, capability, modularity, reliability, interchangeability, efficiency, and cost performance. Attributes are considered to be a combined function of factors. Examples of product factors are market share, flexibility, user acceptance, capacity utilization, safety, and design functionality. Factors are themselves considered to be composed of indicators. Examples of indicators are debt ratio, acquisition volume, product responsiveness, substitutability, lead time, learning curve, and scrap volume. By combining the above definitions, a composite measure of the operational value of a product can be quantitatively assessed. In addition to the quantifiable factors, attributes, and indicators that impinge upon overall project value, the human-based subtle factors should also be included in assessing overall project value. Value is synonymous with quality. Thus, the contemporary earned value technique is relevant for "earned quality" analysis. This is a good analytical technique to use for the justification stage of the DEJI systems model. This will impact cost, quality, and schedule elements of product development with respect to value creation. The technique involves developing important diagnostic values for each schedule activity, work package, or control element. The variables involved are PV: Planned Value; EV: Earned Value; AC: Actual Cost; CV: Cost Variance; SV: Schedule Variance; EAC: Estimate at Completion; BAC: Budget at Completion; and ETC: Estimate to Complete. This analogical relationship is a variable research topic for quality engineering and technology applications.

INTEGRATION OF QUALITY

Integration is the fourth and last phase of the DEJI systems model. Without being integrated, a system will be in isolation and it may be worthless. We must integrate all the elements of a system on the basis of alignment of functional goals. The overlap of systems for integration purposes can conceptually be viewed as projection integrals by considering areas bounded by the common elements of sub-systems. Quantitative metrics can be applied at this stage for effective assessment of the product state. Trade-off analysis is essential in quality integration. Pertinent questions include the following:

What level of trade-offs on the level of quality are tolerable?
What is the incremental cost of higher quality?
What is the marginal value of higher quality?
What is the adverse impact of a decrease in quality?

What is the integration of quality of time? In this respect, an integral of the form below may be suitable for further research:

$$I = \int_{t_1}^{t_2} f(q)\,dq,$$

where I = integrated value of quality, $f(q)$ = functional definition of quality, t_1 = initial time, and t_2 = final time within the planning horizon.

Presented below are guidelines and important questions relevant for quality integration:

- What are the unique characteristics of each component in the integrated system?
- How do the characteristics complement one another?
- What physical interfaces exist among the components?
- What data/information interfaces exist among the components?
- What ideological differences exist among the components?
- What are the data flow requirements for the components?
- What internal and external factors are expected to influence the integrated system?
- What are the relative priorities assigned to each component of the integrated system?
- What are the strengths and weaknesses of the integrated system?
- What resources are needed to keep the integrated system operating satisfactorily?
- Which organizational unit has primary responsibility for the integrated system?

The systems approach of the DEJI model will facilitate a better alignment of product technology with future development and needs. The stages of the model require research for each new product with respect to design, evaluation, justification, and integration. Existing analytical tools and techniques as well as other SE models can be used at each stage of the model. Thus, a hybrid systems modeling is possible.

THE WATERFALL MODEL

The Waterfall model, also known as the linear-sequential life cycle model, breaks down the SE development process into linear sequential phases that do not overlap one another. The model can be viewed as a flow-down approach to engineering development. The Waterfall model assumes that each preceding phase must be completed before the next phase can be initiated. Additionally, each phase is reviewed at the end of its cycle to determine whether or not the project aligns with the project specifications, needs, and requirements. Although the orderly progression of tasks simplifies the development process, the Waterfall model is unable to handle incomplete tasks or changes made later in the life cycle without incurring high costs. This makes sense for the waterfall model since water normally flows downward, unless forced to go upward through a pumping device, which could be an additional cost. Therefore, this model lends itself better to simple projects that are well defined and understood.

THE V-MODEL

The V-model (for verification and validation) is an enhanced version of the Waterfall model that illustrates the various stages of the system life cycle, as shown in Figure 7.2 The V-model is similar to the Waterfall model in that they are

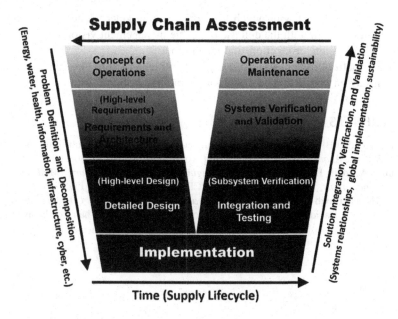

FIGURE 7.2   The V-model of Systems Engineering.

both linear models whereby each phase is verified before moving on to the next phase. Beginning from the left side, the V-model depicts the development actions that flow from concept of operations to the integration and verification activities on the right side of the diagram. With this model, each phase of the life cycle has a corresponding test plan that helps identify errors early in the life cycle, minimize future issues, and verify adherence to project specifications. Thus, the V-model lends itself well to proactive defect testing and tracking. However, a drawback of the V-model is that it is rigid and offers little flexibility to adjust the scope of a project. Not only is it difficult but it is also expensive to reiterate phases within the model. Therefore, the V-model works best for smaller tasks where the project length, scope, and specifications are well defined.

SPIRAL MODEL

The Spiral model is similar to the V-model in that it references many of the same phases as part of its color coordinated slices, which indicate the project's stage of development. This model enables multiple flows through the cycle to build a better understanding of the design requirements and engineering complications. The reiterative design lends itself well to supporting popular model-based SE techniques such as rapid prototyping and quick failure methods. Additionally, the model assumes that each iteration of the spiral will produce new information that will encourage technology maturation, evaluate the project's financial situation, and justify continuity. Consequently, the lessons learned from this method provides a data point with which to improve the product. Generally, the Spiral model meshes well with the defense life cycle management vision and integrates all facets of design, production, and integration.

The benefit of incrementally adding capabilities in a spiral fashion can help a system stay on budget and schedule for operational rollout.

REFERENCES

Badiru, Adedeji B. (2012), "Half-life learning curves in the defense acquisition lifecycle," *Defense Acquisition Research Journal*, Vol. 19, No. 3, pp. 283–308.

Badiru, Adedeji B. (2019), *Systems Engineering Models: Theory, Methods, and Applications*, Taylor & Francis Group/CRC Press, Boca Raton, FL.

Perez, Hernan David (2014), *Supply Chain Roadmap: Aligning Supply Chain with Business Strategy*, David Hernan Perez, Charleston, NC: www.SupplyChainRoadmap.com

Index

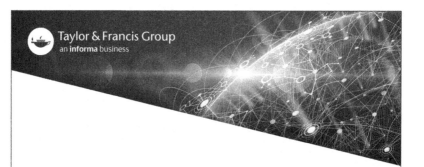

Printed in the United States
by Baker & Taylor Publisher Services